Faculty Health in Academic Medicine

Thomas R. Cole • Thelma Jean Goodrich
Ellen R. Gritz
Editors

Faculty Health in Academic Medicine

Physicians, Scientists, and the Pressures
of Success

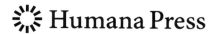 Humana Press

Editors

Thomas R. Cole
McGovern Center for Health, Humanities,
   and the Human Spirit
University of Texas-Houston Medical
   School
Houston, TX

Ellen R. Gritz
Department of Behavioral Science
Olla S. Stribling Distinguished Chair
   for Cancer Research
The University of Texas M. D. Anderson
   Cancer Center
Houston, TX

Thelma Jean Goodrich
Department of Behavioral Science
The University of Texas M. D. Anderson
   Cancer Center
Houston, TX

ISBN 978-1-60327-450-0   e-ISBN 978-1-60327-451-7
DOI 10.1007/978-1-60327-451-7

Library of Congress Control Number: 2008931180

Foreword

In the 21st century, academic medical centers across the United States continue to make scientific breakthroughs, to make improvements in patient care, and to provide the most advanced information and guidance in matters affecting public health. The signs of growth are everywhere—in new research buildings, new partnerships with industry, new forms of molecular medicine, and new sensitivity to the role of the human spirit in healing. This growth is due in large part to the dedication and productivity of our faculty, who are providing more patient care, more research, more teaching, and more community service than ever before.

Today, there are roughly 135,000 physicians, scientists, and other faculty working at approximately 125 academic medical centers around the country. Increasingly, they are asked to do more with less. Since the 1990s, academic medical centers in the United States have lost the financial margin they once enjoyed, thereby putting new pressures on research, education, and clinical care. Medical school faculty, previously given funded time for teaching and research, are increasingly drafted to bring in clinical revenues to cover their salaries. Dedicated to the missions of research, teaching, and care, our faculty have responded well to these challenges and perform at a very high level. However, we are beginning to see the results of ongoing stress.

Recent trends in bioethics have emphasized concern for the patient as a whole person, but very little attention has been paid to the legitimate needs and concerns of physicians, scientists, and other health care professionals. This ground-breaking book is the first to look carefully at issues of faculty health and well-being. It grows out of a conference sponsored by the McGovern Center for Health, Humanities, and the Human Spirit at University of Texas-Houston Health Science Center in the summer of 2007. Its findings and recommendations offer an essential framework for protecting and enhancing the well-being of our faculty, our institutions, and the future of academic medicine.

James T. Willerson
President
University of Texas-Houston Health Science Center
Houston, Texas, USA

Preface

Altruism and self-governance, in addition to an evolving body of knowledge, are among the most important attributes of a learned profession. In medicine, altruism means placing the patient's interests above the physician's interest. Altruism also implies the responsibility of physicians to teach the learned profession to their students. The original Hippocratic Oath specifically identified this responsibility to pass on knowledge and wisdom. Self-governance has traditionally implied physicians' responsibility to be concerned about their colleagues' functioning and quality of care. For centuries, this responsibility took the form of "professional courtesy" in which physicians cared for colleagues, and often their families, without charging a professional fee. In the last part of the 20th century, this tradition gave way to insurance regulations, which precluded its practice in most situations.

Over the past half century, the commitments of the profession to meet the requirements for self-governance and altruism have eroded. Physicians have been extremely reluctant to respond to or identify dysfunctional colleagues. Physicians often fail to intervene when they sense a colleague may have mental health problems. Although over half of medical trainees experience an episode of depression, and a significant number have suicidal thoughts, the stigma associated with mental illness has posed a substantial barrier to students seeking aid. These concerns often prevent students from seeking counseling in institutional facilities unless truly confidential off-campus opportunities are created for the student to seek help.

This culture of denial of mental and physical symptoms among physicians is strongly internalized by students and subsequently becomes an important part of the behavior pattern of health professionals later in their careers. Long working hours, increased pressure to generate income in medical education, accelerating administrative responsibilities, and the shame/blame conundrum in medical education, have served to increase the stresses on health providers, particularly physicians and nurses.

Certain changes in the health care delivery system and in the health professions are beginning to ameliorate these problems. The substantial increase in the number of women in medicine has diminished the role of "machismo" and the "I-can-take-anything-that-the-system-dishes-out" mentality of male trainees. Limitations on the work schedule for residents, the introduction of patient safety, and quality improvement programs that emphasize the analysis of errors rather than blame/shame mentality are all steps in the right direction. At the same time the increasing pressures to be financially productive while teaching—or to be funded for research in an increasing competitive environment—have exacerbated stress. And research scientists, who play such an important role in our academic health centers, are particularly stressed by the increased competition for diminishing federal research dollars.

In this context, altruism and self-governance take on new urgency and new meanings. Self-governance should entail increased attention to promoting wellness and self-care among one's colleagues, especially in our academic health centers. In order to be of maximum service to patients and society, altruism requires that physicians and scientists attend to their own well-being. There is no single explanation for either the high rates of burnout, depression, and suicide among health professionals, particularly physicians, or the psychological dilemmas, which are faced by our clinical and research faculties. What is encouraging, however, is the emerging focus on the importance of faculty health, especially mental health, and the need to better understand the factors which contribute to unhealthy situations.

There is no single answer to the challenge of faculty health. Illuminating the issue, validating its importance, and focusing the intellect of thoughtful individuals on solutions to these problems are important steps forward. In this volume a wide variety of experiences is discussed, and a number of theories are advanced with regard to the faculty health conundrum—from prevention and wellness to diagnosis and treatment. Like many issues in health and science, it is critical that specific hypotheses be advanced and careful efforts made to determine their validity through interventions that are as well controlled as possible. As with all elements of human behavior and human need, these are difficult issues to study. But solving these problems requires that we move from theory to well-constructed research and practice.

This volume offers important opportunities to identify the questions and, in many cases, suggest ways that solutions might be tested. It will be important to share the outcomes of interventions and best practices of the health and science professionals as we attempt to improve faculty health. This is a responsibility academic health centers have to their faculty and faculty in the health professions have to themselves and each other. In the 21st-century academic health center, self-governance requires organizational health and commitment to the well-being of faculty. And altruism—serving others and educating new professionals—requires self-care and care for one's peers.

Kenneth Shine
Executive Vice Chancellor for Health Affairs
Interim Chancellor
University of Texas System
Austin, Texas, USA

Contents

Part VI   Conclusion

Authors and Affiliations

Janis Apted, M.L.S. Faculty Development, University of Texas M. D. Anderson Cancer Center, Houston, TX

Walter Baile, M.D. Department of Behavioral Science, University of Texas M. D. Anderson Cancer Center, Houston, TX

Janet Bickel, M.A. Janet Bickel and Associates, Falls Church, VA

Daria Boccher-Lattimore, Dr.P.H. Clinical Sociomedical Sciences in Psychiatry, Mailman School of Public Health, Columbia University, New York/New Jersey AIDS Education and Training Center, New York

Steven Bogdewic, Ph.D. Department of Family Medicine, Office of Faculty Affairs and Professional Development, Indiana University School of Medicine, Bloomington, IN

Eugene Boisaubin, M.D. Department of Internal Medicine, University of Texas-Houston Medical School, Houston, TX

Nathan Carlin, M.Div. Rice University, Houston, TX

Thomas R. Cole, Ph.D. McGovern Center for Health, Humanities, and the Human Spirit, University of Texas-Houston Medical School, Houston, TX

Elise Cook, M.D. Department of Clinical Cancer Prevention, University of Texas M. D. Anderson Cancer Center, Houston, TX

Ann H. Cottingham, M.A.R. Special Programs, Indiana University School of Medicine, Indianapolis, IN

Richard M. Frankel, Ph.D. Department of Medicine, Department of Geriatrics, Regenstrief Institute, Inc., Indiana University School of Medicine, Richard L. Roudebush VAMC, Indianapolis, IN

Mamta Gautam, M.D., F.R.C.P.C. Department of Psychiatry, University of Ottawa, Ottawa, Ontario, Canada

Harry Gibbs, M.D. Office of Institutional Diversity, University of Texas M. D. Anderson Cancer Center, Houston, TX

Thelma Jean Goodrich, Ph.D. Department of Behavioral Science, University of Texas M. D. Anderson Cancer Center, Houston, TX

Kevin Grigsby, D.S.W. Professor of Psychiatry, Department of Psychiatry, Vice Dean for Faculty and Administrative Affairs, Penn State College of Medicine, Hershey, Pennsylvania, USA

Ellen R. Gritz, Ph.D. Department of Behavioral Science, University of Texas M. D. Anderson Cancer Center, Houston, TX

Jeri Hepworth, Ph.D. Department of Family Medicine, University of Connecticut School of Medicine, Storrs, CT

Richard Holloway, Ph.D. Department of Family and Community Medicine, Office of Student Affairs, Medical College of Wisconsin, Milwaukee, WI

Thomas S. Inui, Sc.M., M.D. Department of Health Services Research, Regenstrief Institute, Inc., Indiana University School of Medicine, Indianapolis, IN

Richard L. Roudebush, V.A.M.C., Indianapolis, IN

Craig Irvine, Ph.D. Department of Medicine, College of Physicians and Surgeons, Columbia University, New York, NY

Debra K. Litzelman, M.A., M.D. Department of Medicine, Indiana University School of Medicine, Indianapolis, IN

John Mendelsohn, M.D. President, University of Texas M. D. Anderson Cancer Center, Houston, TX

Susan McDaniel, Ph.D. Department of Family Medicine, Department of Psychiatry, Wynne Center for Family Research, University of Rochester School of Medicine, Rochester, NY

David M. Mossbarger, M.B.A. Regenstrief Institute, Inc., Indiana University School of Medicine, Indianapolis, IN

Patricia A. Parker, Ph.D. Department of Behavioral Science, University of Texas M. D. Anderson Cancer Center, Houston, TX

Anu Rao, Ph.D. Ombuds Office, University of Texas M. D. Anderson Cancer Center, Houston, TX

Kathleen Sazama, M.D., J.D. Department of Laboratory Medicine, University of Texas M. D. Anderson Cancer Center, Houston, TX

Ken Shine, M.D. Executive Vice-Chancellor for Health Affairs and Interim Chancellor for the University of Texas System, Houston, TX

Henry W. Strobel, Ph.D. Department of Biochemistry and Molecular Biology, Office of Faculty Affairs, University of Texas-Houston Medical School, Houston, TX

Anthony L. Suchman, M.A., M.D. Department of Medicine, Department of Psychiatry, Relationship Centered Health Care, University of Rochester School of Medicine and Dentistry, Rochester, NY

Georgia Thomas, M.D. Employee Health Services, University of Texas M. D. Anderson Cancer Center, Houston, TX

Thomas Viggiano, M.D. Department of Gastroenterology, Office of Faculty Affairs, Mayo Clinic, Rochester, MN

James Willerson, M.D. President, University of Texas-Houston Health Science Center, Houston, TX

Penelope R. Williamson, Sc.D. Department of Medicine, The Johns Hopkins University School of Medicine, Relationship Centered Health Care, Baltimore, MD

Part I
Introduction

Chapter 1
The Context of Concern for Faculty Health

Thelma Jean Goodrich, Thomas R. Cole, and Ellen R. Gritz

> Love is a central theme in the profession of medicine. It is an undeniable drive that leads individuals onward to serve others, whether through clinical care or laboratory science. Naming that force for what it is and claiming it again as one's own turns the corner away from discouragement toward renewal of the promise to that primal altruism which first sparked the notion of joining an honored tradition of service.

So began a three-day working conference on faculty health and well-being in academic medical centers, the first to focus entirely on that subject. Henry Strobel spoke these words at the opening dinner. They both set the theme and recalled the reason for the gathering: faculty burnout and demoralization. The causes are multiple. Sometimes faculty work under onerous conditions—too much to do in too short a time, not enough resources, not enough support staff, and so on. Other times—out of what they regard as dedication and passion—faculty work past needing rest, past needing family, and past needing renewal. Either way, the result is finally a separation from the very inspiration that "sparked the notion of joining an honored tradition of service."

The conference was organized by the editors of this volume and was sponsored by the McGovern Center at the University of Texas Health Science Center in Houston in collaboration with the M. D. Anderson Cancer Center, also in Houston. It had its origins in concern for loss of meaning and its deleterious consequences. But what pulled us forward was envisioning a new field of inquiry that would explore all facets of faculty well-being—the major factors affecting it and resources for protecting, recovering, and enhancing it. Our aim was to convene those with expertise in relevant areas and produce a foundational book for this new field.

T.J. Goodrich
Department of Behavioral Science, University of Texas M. D. Anderson Cancer Center, Houston, Texas, USA
e-mail: tjgoodrich@mdanderson.org

T.R. Cole
McGovern Center for Health, Humanities, and the Human Spirit, University of Texas-Houston Medical School, Houston, Texas, USA

E.R. Gritz
Department of Behavioral Science, University of Texas M. D. Anderson Cancer Center, Houston, Texas, USA

T.R. Cole et al. (eds.) *Faculty Health in Academic Medicine,*
© Humana Press, a part of Springer Science + Business Media, LLC 2009

We set out key areas of discussion:

- Types and prevalence of harms to the health of faculty
- Challenges to health at different stages of the professional life cycle
- Psychological strengths, vulnerabilities, and injuries
- Issues raised by gender, generation, race, and ethnicity
- Helps and hindrances from the professional and organizational culture
- The ethical imperative of self-care
- The spiritual crisis and the role of the humanities
- The need for supportive programs

These topics required experts in epidemiology, impairment, career development, psychology, diversity, organizational phenomena, ethics, medical humanities, and program design. In February, 2007, the organizers identified scholars across the country who had a professional stature in these areas. We wanted to act quickly and set a July date at the University of Texas Health Science Center in Houston. Given this time and place, we were uncertain how many would agree to come. Nevertheless, within a week of our phoned invitations, everyone we had called had agreed to attend. There could not be a stronger endorsement of the importance and urgency of the subject at hand.

Rationale

Undergirding our project was the expanding published research showing that clinicians and researchers in academic medicine, performing daily under high levels of stress, do so at great cost to their health. Many physicians are burned out, demoralized, wounded, and physically compromised [1–3]. Physicians suffer higher levels of anxiety and depression than do those in comparative general populations [4, 5]. A national survey of generalist physicians in the United States found a significant direct relationship between reports of job stress and measures of poorer physical and mental health [6]. A profession rooted in compassion, care, and service to patients has apparently failed to take seriously its own needs for self-care, stress management, meaning, and nurture

Socialized to diagnose and treat disease through biomedical science and technology, physicians sometimes wall themselves off from emotional connection with their patients. As a result, they miss the spiritual sustenance and vitality that emerge from genuine human exchange between a doctor and a patient. Further, in academic medicine, physicians face substantially increased clinical work and are still expected to participate in teaching, research, service to the university, and writing for publication—activities that bring their own stresses and create constant conflict about use of time.

Researchers, too, suffer from stressors endemic to scientific studies in an academic medical setting. Most often, they rely on grant support for their research and all or most of their salary. Submissions are not only highly competitive for the initial award of a project, but also for its continuation. Recently, the overall success rate has fallen to about 20% of grant applications submitted to the National Institutes of Health [7]. In 2005, only 9% of all investigators secured independent investigator-initiated (R01) funding on the first round, an outcome necessitating multiple submissions [8]. Finally, in the past few years, awarded grants have been cut 15–30% from the original budget levels [9]. Junior investigators are at greatest risk of being driven out of the field, and even seasoned investigators are feeling the strain severely.

In sum, academic researchers function at the pressure point between the scientific importance of their work and its unstable future. Additionally, they must fund their research staff, protect sought-after lab space, publish steadily, teach, have national visibility, and provide both intramural and extramural service. Thus, like physicians, they feel the constant pull of attending to multiple masters. Health does not thrive under these pressures, nor do family and personal relationships.

An academic medical center's lack of attention to human resources is not only short-sighted, it is expensive as well. The costs of faculty turnover are estimated to be 5% of a center's budget, not including the costs of lost opportunities, lost referrals, overload on remaining faculty, reduced productivity, and lower morale. Likewise, staff turnover is not only costly in itself, but also stresses faculty and disrupts their productivity.

Defining the Territory

The conference was divided into four sections: examination of faculty health, personal and social dimensions, perspectives from the humanities and social sciences, and supports and interventions. Several authors contributed chapters for each section. What follows is a summary of key findings, concerns, and recommendations.

Examination of Faculty Health

Epidemiology. As detailed in Chapter 2 by Daria Boccher-Lattimore, data regarding morbidity, mortality, and health care provide a telling perspective on the state of health of academic faculty in medicine. Because virtually all the surveys focus only on physicians—and many of those surveys target physicians working in the community—not enough is known about the health of academicians, whether researchers or physicians. This gap stands as a major area to fill with upcoming studies.

Impairment. Special attention is given in most settings to what is termed the "impaired physician." Chapter 3 by Eugene Boisaubin gives the definition of impaired physician held by The American Medical Association: one who is "unable to fulfill professional or personal responsibilities because of a psychiatric illness, alcoholism, or drug dependency." Dr. Boisaubin proposes that the list ought to include chronic, unremitting stress because it too can affect judgment and performance.

Methods and uses of measuring faculty health. Mamta Gautam in Chapter 4 suggests a number of measures of wellness that can be used in the workplace to determine the presence and level of burnout and illness. The results of such measures can lend support for the creation of a wellness program for the faculty. Specific steps are outlined to facilitate the development of such a program.

Personal and Social Dimensions

Psychological health. The interplay between the academic medical center and its faculty shapes the health of both. Susan McDaniel, Stephen Bogdewic, Richard Holloway, and Jeri Hepworth advise, in Chapter 5, that leaders must have the vision and the courage to align systems in a manner that ensures both individual and organizational success. In turn, faculty must assess their own talents and interests and determine how they align with organizational goals and priorities. The authors provide tables offering a framework for assessing the psychological health of the individual and the medical center.

Faculty life cycle. Stage of career is relevant for understanding and responding to changing stressors and needs of faculty. In Chapter 6, Thomas Viggiano and Henry W. Strobel present The Career Management Life Cycle Model. It identifies eight phases: recruitment, orientation, exploration, engagement, development, vitality, transition, and retirement. Each phase provides the institution with an opportunity to give targeted and tailored support to assist the individual. In many instances, however, the culture of the institution requires significant change before any efforts to help the individual can be successful.

Gender and generation. Among the complex factors shaping the experience of and response to the exacting conditions of academic medicine, gender and generation stand at the forefront. In Chapter 7, Janet Bickel emphasizes both the work that remains to facilitate women realizing their potential and the newer challenge of bridging generational differences. In light of the resulting demands placed on senior faculty, she outlines promising directions for leaders in academic medical centers who are forward-looking enough to place faculty vitality high on their list of priorities.

Diversity. Faculty members who belong to underrepresented minorities have unique stressors in addition to those shared with their colleagues. Elise Cook and Harry Gibbs specify in Chapter 8 that marginalization, lack of mentoring, limited networking opportunities, social isolation, devaluation of their work, and lower rates of promotion than those of the majority of the faculty are marked examples. The authors examine the consequences of discrimination and other stressors not only for careers, but also for morale and health.

Perspectives from the Humanities and Interpretive Social Science

Organizational culture. Each academic medical center has its particular organizational culture, or shared pattern of basic assumptions about "how things are done around here." In Chapter 9, Kevin Grigsby describes some organizational cultures as conflict laden and competitive, while others value nurturing and mentoring. More explicit attention to faculty health and wellness requires not only programs of prevention and enhancement, but also changes in organizational culture needed to promote faculty well-being.

The ethics of self-care. The medical academy's primary ethical imperative may be to care for others, but this imperative is meaningless if it is divorced from the imperative to care for oneself. So argues Craig Irvine in Chapter 10. How can we hope to care for others, after all, if we, ourselves, are crippled by ill health, burnout, or resentment? Too often, however, this imperative remains unheeded by medical academicians and ignored by professional ethicists. If they are to heed the self-care imperative, medical academicians must turn to an ethics that not only encourages, but even demands care of the self. An important resource can be found in narrative ethics. Since narrative is central to the understanding, creation, and recreation of our selves, we can truly *care* for our selves only by attending to our self-creating stories. Narrative ethics brings these stories to our attention; so doing, it allows us to honor the self-care imperative.

The humanities. From the perspective of scholarship in the humanities, faculty health is closely tied to the question of meaning. In Chapter 11, Thomas Cole and Nate Carlin explain that this focus is stated clearly in the definition of health put forward by the AAMC task force in 1999: health is "not just the absence of disease but a state of well-being that includes a sense that life has purpose and meaning." Academic physicians and scientists come to their work motivated by the values of science, compassionate care, service, and education. Yet current conditions (especially the requirement to produce clinical income or grant funds) often limit faculty's ability to live up to their highest ideals of service and teaching, a failing which creates cognitive dissonance. Cognitive dissonance in turn may lead to cynicism, disillusionment, self-doubt, disease, and retreat from those ideals that now seem so unrealistic. The humanities cannot "solve" these problems, but they can help faculty and institutions understand and address them.

Reclaiming the call. In Chapter 12, Henry W. Strobel reminds us that since antiquity, medicine has been a calling—a vocation dedicated to comforting, caring, and curing. Yet contemporary financial pressures and challenges of the health care delivery system combine to inhibit one's ability to give to others, to enter into a genuine relationship with patients. Medicine as a calling is undermined, a situation resulting in a loss of heart in faculty and a loss of faculty in institutions. The health and well-being of faculty and of institutions can be enhanced by making personal and structural changes designed to connect faculty with meaning in work.

Supports and Interventions

Developing a faculty health program. Ellen R. Gritz, Janis Apted, Walter Baile, Kathleen Sazama, and Georgia Thomas describe in Chapter 13 the beginnings and subsequent growth of the faculty health program at the University of Texas

M. D. Anderson Cancer Center in Houston, Texas. The program consists of prevention, intervention, and response. The preventive aspect offers seminars on stress, burn-out, resilience, human performance, and productivity, as well as skill-building in meditation, relaxation, and mental fitness. Customized programming is given to departments, chairs, and other faculty leaders, and faculty spouses. A Faculty Assistance Program offers confidential psychological consultations off-site for faculty and their families at no cost to them. The intervention and response programs put an institutional plan in place when an emergency or death occurs among the faculty. Faculty health initiatives are enhanced by collaborations with other institutional programs such as the comprehensive Faculty Development Program, a Faculty Leadership Academy, a Women Faculty Program, an Office of Institutional Diversity, an Ombuds Office, and an I*CARE program (Interpersonal Communication and Relationship Enhancement).

Changing the culture in academic medicine. Establishing a better environment for the challenging work of academic medicine presents itself as a more efficient way to aid faculty members than only targeting the people themselves. In Chapter 14, Debra K. Litzelman and her colleagues recount a unique initiative at the Indiana University School of Medicine that affected faculty well-being through an effort at comprehensive cultural change. The initiative was based on applying relationship-centered care not only to doctors and patients, but also to all members of the academic community. Early efforts at cultural change focused on the formal curriculum and on creating a broadly distributed written document regarding the organization's guiding professional values. Over a several year period, a wide variety of programs were offered regarding personal formation (knowing self), community formation (finding community), and cultural formation (creating value).

Conflict resolution. Several factors typical of academic medical centers contribute to significant potential for conflict among faculty. Examples of such factors include competition for resources, financial strains, turnover in leadership, and a negative climate for funding research. In Chapter 15, Anu Rao, Patricia Parker, and Walter Baile review strategies for managing conflict in academic medical centers. In addition, the authors present an approach developed at The University of Texas M.D. Anderson Cancer Center that utilizes an organizational ombudsperson.

Programming for faculty health and well-being. Most institutions have not initiated projects to change their culture in a substantial way, but some have initiated programs to help faculty members gain knowledge, skills, and opportunity to support self-care. That effort itself frequently stands as strong indication of care on the part of the institution. Faculty health programs may also gather data about sources of stress, sources that leaders may then work to resolve at the institutional level. Programming generally includes a range of offerings such as educational seminars on methods of reducing stressful reactions, confidential psychological counseling, meditation groups, and assessment tools and courses through a web site. Individual departments may also take responsibility for providing programs aimed at promoting stress reduction or increasing morale quite apart from what the formal faculty health program offers. Indeed, surveying the faculty either by department or throughout the institution for their interests and needs creates a strong support and guide for programming.

Setting Forth

The foregoing summaries can only hint at the richness of ideas, innovations, theories, and needs given in the full chapters ahead. It is hoped that reading them will generate for the reader many new and interesting perspectives. These can then lead to contributions not possible before this conversation was engaged.

References

1. Adams, D. (2007) Doctor morale shaky as practice stressors surge (http:www.ama-assn.org/amednews) January 15, 2007.
2. Spickard, A., Gabbe, S.G., and Christensen, J.F. (2002) Mid-career burnout in generalist and specialist physicians *Journal of the American Medical Association* **288(2)**, 1447–1450.
3. Schindler, B.A., Novack, D.H., Cohen, D.G., Yager, J., Wang, D., Shaheen, N.J., Guze, P., Wilkerson, L., and Drossman, D. (2006) The impact of the changing health care environment on the health and well-being of faculty at four medical schools *Academic Medicine* **81**, 27–34
4. Aasland, O.G., Olff, M., Falkum, E., Schweder, T., and Ursin, H. (1997) Health complaints and job stress in Norwegian physicians: the use of an overlapping questionnaire design *Social Science & Medicine* **45(11)**, 1615–29.
5. Sutherland, V.J., and Cooper, C.L. (1992) Job stress, satisfaction, and mental health among general practitioners before and after introduction of new contract *BMJ* **304**, 1545–48.
6. Williams, E.S., Konrad, T.R., Linzer, M., McMurray, J., Pathman, D.E., Gerrity, M., Schwartz, M.D., Scheckler, W.E., and Douglas, J. (2002) Physician, practice, and patient characteristics related to primary care physician physical and mental health: results from the Physician Worklife Survey *Health Services Research* **37(1)**, 121–43.
7. The Cancer Letter (February 8, 2008) **34(5)**, 1–4.
8. Mandel, H.G., and Vesell, E.S. (2006) Declines in funding of NIH R01 Research Grants Letters *Science* **313**, 1387.
9. NCI Funding Policy For FY 2008 Research Project Grant (Rpg) Awards. National Cancer Institute. 14 February 2008. http://deainfo.nci.nih.gov/grantspolicies/FinalFundLtr.htm

Part II
Examination of Faculty Health

Chapter 2
Epidemiology

Daria Boccher-Lattimore

Abstract In this chapter we will review what is known about the health of academic medical faculty and the related morbidities and mortalities. How are these experiences affected by gender, ethnicity, age, and cohort? What are the implications of an unwell academic faculty workforce? Other than the expected effect of impairment and performance, recent studies have shown the relationship between physician job dissatisfaction and less than optimal patient care. These results will be summarized and highlight the personal, professional, and institutional impact of faculty health.

Keywords Academic medical faculty, health status, mortality, morbidity, barriers to care

Introduction

17,596
the # of published articles with "health status" as a key subject heading
3,502
the # of published articles with "medical faculty" as a key subject heading
0
the # of published articles with "health status" and "medical faculty" as key subject headings[1]

D. Baccher-Lattimore
Assistant Professor of Clinical Sociomedical Sciences in Psychiatry, College of Physicians and Surgeons, and Mailman School of Public Health, Columbia University
Director, New York/New Jersey AIDS Education & Training Center, New York, USA
e-mail: dmb82@columbia.edu

---

[1] Based on Ovid Medline MeSH search for journal articles published 1950 through January 2008

T.R. Cole et al. (eds.) *Faculty Health in Academic Medicine,*
© Humana Press, a part of Springer Science + Business Media, LLC 2009

"Faculty health" is an elusive concept. While on the surface a seemingly obvious notion, defining it in an inclusive and measurable way becomes challenging. Who are medical and scientific faculty? And how do we define their health? Once one answers these questions and identifies an operational definition, the challenge becomes identifying available measures, as demonstrated in the example above. Although there is a significant amount of published literature on subjects such as stress and burnout among physicians (but not on scientists), this chapter addresses the dimensions of faculty health more broadly.

There are nearly 125,000 full-time medical faculty in the United States, including clinical specialists and generalists, and basic, behavioral, and social scientists [1]. The majority (62%) are medical doctors; a quarter of them are Ph.D. or other doctorate-level faculty; and 7% have dual degrees (M.D./Ph.D.). In order to meet the institutional missions of the academic medical center, faculty divide their time among research, teaching, and clinical duties to varying degrees. Recent changes in the environment of academic health care settings, decreasing financial security, and increasing demands on faculty have created an increasingly stressful environment for the academic medical faculty member. Stress and workplace factors have long been known to have ill effects on the workforce. However, little is known if and how the changing academic environment has impacted the health of its faculty.

The definition of health has evolved from the biomedical definition of "the absence of disease" to a multidimensional concept, which includes physical, social, mental, and spiritual well-being. Indeed, some have argued that measures of signs and symptoms of disease are not sufficient measures on health, rather functional outcomes are necessary to truly understand a population's health [2]. Yet as our conceptualization of health has expanded, the availability of adequate measures of such has lagged behind, particularly, as we seek to assess the health of a population. Our most readily available indicators of a population's health remain measures of disease and its consequences, i.e., morbidity and mortality.

Indicators of a population's health tend to come from three sources: vital statistics, surveys/self-reports, and information on health services utilization. Vital statistics provide us with counts on a population's mortality and the incidence and prevalence of some diseases. This information is often supplemented with survey data, which range from large surveys of nationally representative samples to cohort studies focusing on small well-defined populations. These surveys allow for measures of experiences of distress and functioning and their impact on quality of life. Finally, increasingly, researchers have turned to measures of health care utilization and health practices as indicators of a population's health.

This chapter will review the empirical literature on the mortality, morbidity, and health practices of academic medical faculty.

Mortality

14,008
the # of published articles with "mortality" as a key subject heading

3,502
the # of published articles with "faculty, medical," as a key subject heading
0
the # of published articles with "mortality" and "medical faculty" as key subject headings[2]

While mortality data are gathered on a regular basis, the literature is sparse on the mortality of academic medical faculty. Tens of thousands of articles address mortality and thousands are available focusing on medical faculty, yet no studies were referenced in the medical literature by both keywords, "mortality" and "faculty, medical". This is a function of the sources of mortality data and the available measures of population subgroups. Larger vital statistic databases are the primary source of mortality data. While these generally include measures of occupation, they do so with generic (often census-based) categories. These data allow for, at best, a comparison of similar occupational categories, e.g., physicians and other professionals. Further differentiation of occupational group is not possible; so, for instance, one cannot distinguish between physicians practicing in the community from those based in an academic medical center. Cohort studies are another source of cause-specific mortality data, allowing for a focused analysis of a well-defined population. However, these studies often lack external validity, i.e., have limited generalizability to groups outside of the study population. In addition, mortality data of physicians and academic medical faculty is limited by the fact that women and minorities entered this workforce relatively recently; hence, available data are insufficient for subgroup analysis.

One of the few comprehensive studies on all-cause and cause-specific mortality among physicians was conducted by Frank et al. with a proportionate analysis of data from the National Occupational Mortality Surveillance database [3]. The National Institute for Occupational Safety and Health (NIOSH) maintains this database of death certificate data with occupation information. The usual occupation of the decedent is coded according to the Bureau of the Census classification system, allowing for comparisons of mortality across similar job classifications.

The authors compared the proportion of deaths due to a specific cause in physicians with the proportion of that cause of death in lawyers and all professionals. The cause of death of physicians and other professionals over the age of 25 reported in 28 states between 1984 and 1995 were the basis of analysis. Analyses were gender- and race-specific and were limited to those with a race/ethnicity of white or black, due to small numbers of other races. However, data for women were not presented because there were relatively fewer older women in the physician population.

Nearly four million deaths were reported of men aged 18–90 years and whose race was either black or white: 204,365 white male professionals, 13,034 white male physicians; 13,558 professional black males, and 347 black male physicians.

---

[2] Based on Ovid Medline MeSH search for journal articles published 1950 through January 2008

Overall, white male physicians lived longer (73 years, mean age at death) than lawyers (72.3 years), other professionals (70.9 years), and men in the general population (70.3 years). A similar result was found among black males. Black male physicians lived longer (68.7 years) than other professionals (65.3 years), men in the general population (63.6 years) and lawyers (62 years). (Curiously, black male lawyers had the youngest mean age of death.) Stark, is the racial disparity in mean age of death, even among professionals. Black male physicians, while having the highest mean age at death among black males, had a younger mean age of death than all of the white male categories.

The overall finding that physicians live longer than others is not unexpected given the high socioeconomic status associated with the profession. However, this holds true even when comparing to professionals of assumed similar socioeconomic status. The reasons for this are not known, but may be attributed to better access to health care, more awareness of healthy behaviors and/or health practices. These will be examined below.

Frank et al. examined cause-specific mortality as well. Such subgroup analyses further reduced the sample size, so that cause-specific analyses concentrated on that of the white male physician populations. In general, the most common causes of death for white male physicians were similar to those in the general population: heart diseases and cancers. Elevated rates of death among white male physicians were found with accidents and suicide and drug-related causes. A similar analysis of the limited number of white female physicians shows an elevated rate of death due to drug-related deaths, suicide, and self-inflicted injuries. Black male physicians only differed from other black male professionals in a higher rate of diabetes mellitus-related deaths (but again the numbers were very small in these subgroup analyses).

In a study to determine the relationship between occupation and death from ischemic heart disease (IHD) (using the same database as Frank above), Calvert et al. (1999) found racial differences for physicians as well [4]. In white-collar occupations for black males, physicians had the highest proportionate mortality risk from IHD for all professions; while white male physicians had a significantly lower risk than their white-collar counterparts. These data further support the preliminary findings that racial differences may exist in cause-specific mortality among physicians.

Three earlier cohort studies support the conclusion that, in general, physicians experience a lower cumulative mortality than their counterparts in the general population [5–7]. Williams et al. (1971), in the earliest of these studies, reported that the expectation at the time was, in fact, that physicians would experience an elevated level of mortality than the general public due to higher levels of stress and longer working hours. Two of these cohort studies examined cause-specific mortality [5]. In a cohort study of medical school graduates from two California schools, Ullman et al. (1991) found elevated risks of suicide among one cohort and elevated risk of death by accidents in both cohorts, compared to the general US white male population [6]. In an analysis of young physician deaths reported in JAMA over an eight-year period, Sankoff et al. found an overall favorable mortality rate of the physicians, as compared to the general population; however, an elevated risk of preventable deaths (suicide, homicide, and unintentional injuries) [7].

There are limitations to these studies, in that they were conducted on small cohorts and caution is suggested in interpreting these results beyond the population of investigation. Indeed when analyzing small cohorts it is often difficult obtain an adequate comparison group. However, the pattern of these findings, i.e., elevated risk of suicide and accidental deaths among physicians at the least warrants further investigation.

In a comprehensive systematic review of published epidemiological studies on suicide and physicians, Lindeman et al. found an elevated risk of suicidal death for male physicians ranging from a 10% increase to over triple the rate; for female physicians, the rates were more than double to over five times that of the general population [8]. When compared with other professionals, male physicians were also found to have an elevated risk of suicide ranging from 1.5 to 3.8 times and for females, 3.7 to 4.5 times.

In response to these reports, in 2002, the American Foundation for Suicide Prevention convened a workshop to develop a consensus statement on what is known about the prevention of suicide in physicians. The resultant report was published in the Journal of the American Medical Association [9]. After a review of the literature, the group identified areas of further research and recommendations for reform. The planning group recognized the need to address the culture of medicine, which, it concluded, despite the burden of these findings, was slow to prioritize the mental health needs of physicians. There were substantial barriers to seeking mental health care, including those that may ultimately impact the livelihood of the physician, i.e., licensure and privilege consequences. As a result, recommendations were made at both the individual and institutional levels to address professional attitudes and institutional policy in order to encourage physicians to seek mental health care.

Since the release of the consensus statement, further studies have confirmed the elevated risk of suicide among physicians. Schernhammer and colleagues conducted a meta-analysis of published literature on suicide in physician populations, to quantitatively assess the findings across studies [10]. They found 25 datasets on physician suicide rates from articles published between 1960 and July 2003, which met their entry criteria. Male physicians experienced an elevated risk of suicidal deaths (40% higher) than males in the general population. The effect for female physicians was more than double the risk of the general population, with the earliest studies reflecting an even higher elevation of risk for female physicians. These results not only confirm elevated risks of suicide among physicians, but highlight a significant difference by gender.

In sum, generally, physicians experience an overall advantage in mortality rates compared to other professionals and the general population. This appears to hold true for both white and black physicians, when compared to race-specific referent groups. However, the racial disparities, i.e., higher mortality rates among blacks, evident in the general population appear to be present as well. It is not clear whether or how these disparities hold for minority female physicians. Further analysis of these differences and their underlying causes is needed.

The cause-specific analyses also point to an elevated risk of mortality due to preventable causes, particularly suicide and accidents among physicians as a whole. The absence of information on physician subgroups (gender, race, practice type, or primary affiliation for example) prevents us from understanding the differences (and/or similarities) among physicians. Indeed, the complete focus on physicians in these studies also leaves us with the larger question whether such is true of all academic medical faculty. How do these experiences in mortality relate to other faculty members, including basic, behavioral, and social scientists? These populations share some of the same academic pressures; further research regarding their experiences of mortality is needed.

Given the elevated risks of preventable mortality one needs to consider the underlying morbidities, health behaviors, and utilization of physical and mental health services of physicians.

Morbidity

5,446

the # of published articles with "morbidity" as a key subject heading

3,502

the # of published articles with "medical faculty" as a key subject heading

2

the # of published articles with "morbidity" and "medical faculty" as key subject headings[3]

Unfortunately, the literature on the larger experiences of morbidity among academic medical faculty includes a scant two studies [11, 12].

The literature documenting the morbidity of physicians and academic faculty members is dominated by that of symptom expression, and experiences of distress and burnout. Recent studies have shown that 30–60% of physicians report experiences of distress and burnout, with an increasing prevalence among younger physicians. (A fuller discussion of the experience of burnout and its consequences is found in Chapter 3.) The implications of these experiences of high levels of stress are frequent reports of health complaints and higher levels of anxiety and depression among physician populations than in the comparative general populations [13, 14].

In one of the earliest studies of its kind, Linn and colleagues (1985) conducted a survey of full-time academic and practicing clinical faculty affiliated with a Department of Medicine at a large Californian urban medical school to examine the relationship between stress, job satisfaction, and health [12]. Questions included general demographics, practice setting descriptors, and how professional time was

---

[3] Based on Ovid Medline MeSH search for journal articles published 1950 through January 2008

spent. Measures of job satisfaction, life satisfaction, depression, anxiety, work/ social conflict, and chronic disease symptoms were the central outcomes. The 50% random sample of faculty was composed entirely of physicians. Academic faculty worked longer hours and took less vacation time than their clinical peers. Clinical faculty devoted almost all their time to direct patient care and administrative activities. Academic faculty spent nearly equal time in patient care, research, and supervising house staff, while administrative activities and teaching rounded out a work week. The obtained measures of mental health revealed that 10% had scores indicative of mild depression and 4% with scores of moderate or marked depression. These were levels comparable to those found in the general population at the time using the same scale [12]. However, 27% had anxiety scores above normal (14% mild anxiety, 13% moderate to severe anxiety) using the Rand Anxiety scale. This was twice the prevalence of depression in this population. No differences were found between the academic and clinical faculty.

Job satisfaction measures indicated that academic and clinical faculty were equally satisfied with their jobs, but derived satisfaction from different sources. For example, academic faculty were more satisfied with their collegial relationships with other disciplines, while clinical faculty were more satisfied with the manpower resources available to them. However, academic faculty reported significantly higher levels of role strain compared to the clinical faculty. Academic faculty scored higher on a scale of work/social conflict, which included items such as feeling torn between demands of work and personal life. Clinical faculty experienced more recent physical symptoms than academic faculty. It appears that the differences in the role demands are reflected in their experiences of stressors.

Schindler et al. (2006) modified the survey used by Linn to survey full-time academic physicians and basic science faculty at four US medical schools in the east, southwest, and western United States to assess whether work-related stressors negatively affected physical and mental health [15]. Academic faculty reported a greater prevalence of symptoms consistent with clinical depression than in a comparable nonpatient population. Using the Center for Epidemiology Study Depression scale, they found elevated experiences of depression symptomatology among both male and female faculty members (20% and 22%, respectively). However, when asked if they experienced "depression" in the past five years women faculty were nearly twice as likely as men to report such (21% and 11%, respectively). While this is closer to the national averages of depression diagnosis by gender [16], it is not clear whether this reflects a true experience of depression, the greater likelihood that female faculty are to report experiencing depressive episodes, or the greater likelihood that female faculty have received a diagnosis of depression. Regardless, the elevated experience of depressive symptoms (nearly double that expected for male faculty as compared to general population) warrants further attention (see Chapter 3).

Schindler et al. also found similar levels of anxiety among physicians as did Linn years earlier; 28% had anxiety scores above normal, with women reporting slightly higher rates of anxiety [15]. Younger faculty, however, reported higher levels of both depression and anxiety than did older faculty members. It is not clear whether the prevalence of increased depressive symptomatology in the younger cohort is a reflection of a true increase or bias due to older faculty with depressive symptoms dropping out of academic life, thereby reducing prevalence in the older cohort. In addition, Schindler's cohort also scored lower on all satisfaction scales than did Linn's sample nearly 20 years earlier.

This growing level of discontent among faculty members and the elevated level of depressive symptomatology in the cohort are troublesome. High levels of depressive symptoms have also been reported in other cohorts of academic physicians. Reinhardt and colleagues in a study of physicians at one academic medical center, similarly found elevated levels of depressive symptomatology among faculty and house staff [17]. Nearly 30% reported depressive symptoms, with a higher rate among younger faculty and a slightly higher rate for women.

The Johns Hopkins Precursors Study is a prospective study of Johns Hopkins medical school graduates from the classes of 1948 to 1964. The participants completed an annual survey since the beginning of this study in 1946. These data have yielded some important results, including a study linking depression and heart disease [18]. After 40 years of follow-up the cumulative incidence of clinical depression was 12%, 132 of 1,190 responding men reporting the development of clinical depression.

Williams et al. report, on the Physician Worklife Study, a survey of clinically active primary care physicians in the United States [19]. This study investigated the impact physician, practice, and patient characteristics have on stress and job satisfaction and, in turn, on reports of general mental and physical health. The authors draw on the conceptual models of stress and coping of Lazarus and Folkman and that of stress and the work of Ivancevich and Matteson [20, 21]. The model proposes three sets of stressors (physician, practice, and patient characteristics) that result in a cognitive appraisal of stress, these antecedents and stress affect job satisfaction. Together, perceived stress job satisfaction affect perceptions of physical and mental health. Structural equation modeling was used to assess these relationships. The survey was distributed to a stratified random sample using the AMA master file as a sampling frame to insure a respondent pool with sufficient diversity in demographic and work characteristics. The sample of clinically active primary care providers was 69% male and 65% white with an average age of 47 years.

Physician and practice characteristics were found to predict job satisfaction, while practice characteristics were the only significant predictors of perceived stress. Specifically, feelings of control over workplace and administrative issues, as well as, balance of work and family life were negatively associated with perceptions of stress. As far as their outcome measures, job stress did predict job satisfaction. While job satisfaction was a significant predictor of positive mental health, perceived stress was a stronger predictor of both poorer reported physical and mental health. The authors conclude that their model confirms the proposition that the practice environment does impact physicians' perception of stress and job satisfaction and consequently physical and mental health. While the sample was limited to providers who devote most of their time to clinical care, it does provide a powerful model of how the practice environment can impact physician health.

In sum, further research is needed to understand the experiences of morbidity on the academic medical faculty as a whole and how they may be different, or the same, within subgroups. The consistent findings of elevated levels of anxiety and depressive symptomatology in combination with that of an elevated risk of mortality due to suicide, in the physician population, cannot be ignored. Indeed these findings point to the imminent need for well-designed health interventions. The work of Williams et al. [19] delineating the interpersonal and environmental factors (including the practice environment) that impact physician physical and mental health provide direction for these interventions (see Chapters 13–16).

Health Utilization and Behaviors

9,171
the # of published articles with "health behaviors" as a key subject heading
3,502
the # of published articles with "medical faculty" as a key subject heading
4
the # of published articles with "health behaviors" and "medical faculty" as key subject headings[4]

Given the intrinsic access to health services among medical faculty, one might expect an appropriate, if not high, level of service utilization. Studies have consistently reported that in fact, physicians and other academic faculty are underutilizers of formal health care and report foregoing care even when needed.

Kahn et al. in the earliest of these studies (and one of the four that surfaced in the keyword search above) assessed the health maintenance attitudes and behaviors of physician faculty at one medical school and nonphysician faculty from two neighboring colleges [22]. (The authors cite the low numbers of non-M.D. faculty at the medical school as a reason to seek faculty at neighboring institutions.) This is one of the rare studies that focused on university-based physician- and doctoral-level faculty and not the general population of physicians.

Physician faculty members were significantly less likely to report having a primary care provider (44% of M.Ds. vs. 74% of non-M.Ds.), although this difference ceased to exist in older faculty (over 50 years). The nonphysicians were more likely to report that having a health maintenance exam is important and more likely to have had a health maintenance exam in the previous two years (33% M.Ds. vs. 56% non-M.Ds.). Nearly half of the M.Ds. (45%) and 21% of the non-M.Ds. reported never seeing a physician for health maintenance. Both groups reported the need for health maintenance visits more often than they had actually received health maintenance care. The authors suggest several barriers for physicians formally seeking

---

[4] Based on Ovid Medline MeSH search for journal articles published 1950 through January 2008

care, among them concerns of confidentiality and the potential that some physicians may be providing their own care.

Gross et al. followed a cohort of Johns Hopkins School of Medicine graduates from 1948 through 1964 to find predictors of having a regular source of care. Overall, 34% had no regular source of care [23]. Neither age nor gender was significantly associated with having a regular source of care; however, specialty was, ranging from 46% of pathologists to 21% of psychiatrists having no regular source of care. This study was unique in that it assessed the relationship of health locus of control (beliefs that one's health is mainly influenced by internal factors, i.e., ones own actions and external factors, such as powerful others, i.e., health professionals or chance). Those scoring high in the chance domain, as well as, those scoring high in the internal domain were more likely to not have a regular source of care. They also found having a regular source of care correlated with having had six different preventive health services as recommended by USPSTF guidelines. This held true even after controlling for age, sex, and health status.

Wachtel et al., surveying a sample of Rhode Island physicians, found the overall use of formal health services to be very low; their number of office visits was a fourth of the national average. Two thirds of the respondents reported having a primary care physician; however, 12% reported that it was themselves and 30% their partner [24].

The findings in multiple studies of the lack of a source of regular care most likely point to the fact that physicians either rely on self-care or informal care mechanisms. Therefore, the number of visits or the perception of having or needing a regular source of care may be less than expected.

Given the prevalence of anxiety and depressive symptoms in the physician population, it begs the question whether underutilization of health services and self-care is true for mental health services as well. The utilization of mental health services among the general population in the United States is reportedly low. Most people in the United States with mental disorders remain untreated or undertreated. The National Comorbidity Survey-Replication was a household survey of a nationally representative sample using a fully structured diagnostic interview conducted from 2001 to 2003, to examine the prevalence of mental health morbidities and treatment received. Of respondents with a mood, impulse control, anxiety, and/or substance disorders episode within the previous 12 months, less than half (41%) received some treatment [25]. Studies on medical faculty use of mental health services are scarce; studies on student and resident utilization of mental health services show that only 22% of those expressing depressive symptoms actually seek care [26]. The same study showed concerns of confidentiality and impact on career as major barriers to seeking care.

The reliance on self-treatment is also worrisome when it comes to prescription medication. Christe et al. studied the source of care and self-prescribing behaviors of resident physicians in four internal residency programs in the United States [27]. About half indicated that they had no primary care physician or served as their own physician. In addition, respondents were asked whether they used a prescription drug during their residency and if so, how it was obtained. Of those who had taken prescribed drugs over half (52%) admitted to self-prescribing and another 6% received prescriptions from a fellow resident. The source for prescription drugs for those who self-prescribed was most frequently reported to be the "sample cabinet" (42% of self-prescribed drugs and 36% of all meds taken).

Larger studies on physician health behavior are needed to assess how physicians obtain needed mental and physical services and what effect self-care and self-prescribing has on health outcomes.

Two surveys, the Women's Physician Health Study and the Physicians Health Study (PHS), are among the largest surveys of physician populations in this country. The Physicians Health Study (PHS) was a randomized, double-blind, placebo-controlled trial designed to test the effects of low-dose aspirin and β-carotene on cardiovascular disease and cancer among US male physicians [28]. In the early 1980s, the study enrolled male physicians between 40 and 84 years of age, who lived in the United States and who were members of the American Medical Association, without a history of myocardial infarction, stroke, or transient ischemic attack; cancer; current renal or liver disease; peptic ulcer; gout; or contraindication to or current use of either aspirin or β-carotene. The PHS has become a great source of data on a cohort of physicians and over 300 publications have been published from this survey. The cohort consisted of 22,071 physicians; their mean was age 53 years and they represented all geographic regions of the United States. However, since the focus of the study was to investigate the relationships between specific causative factors and disease outcomes, this was a relatively healthy group of physicians and the clinical data collected was specific to the diseases and causes of interest. Baseline data do provide information on health practices of this cohort: 11% were current smokers, 25% drank on a daily basis, 13–14% reported being overweight, and exercised on average 10 days/month [28].

The Women Physician's Health Study (WPHS), on the other hand, was specifically designed to assess health behaviors and counseling practices of US women physicians [29]. Conducted in the early 1990s, it was the first comprehensive survey of its kind. The sampling frame was derived from the AMA Physician master file and a stratified sample was obtained to represent medical school graduates from the four decades of 1950–1989. Frank et al. [29] in analyses of WPHS and data from the 1992 Behavioral Risk Factor Surveillance Survey (BRFSS) of the Centers for Disease Control and Prevention compared health-related behaviors of women physicians to women of high socioeconomic (SES) status and women in the general public.

Looking at seven recommended screening behaviors, the physicians received screening for the most part as frequently as those of women in the high SES group, both of which were more likely to have been completed and more recently than women of lower SES in the general population. With two exceptions, physicians were less likely to have had a recent pap test than their high SES counterparts and older women physicians were less likely to have had a recent breast exam by a clinician than their high SES counterparts, perhaps indicative of the lack of regular source of care and the reliance on self-care and informal mechanisms.

Personal Behaviors

The most frequent behaviors investigated among physician populations are alcohol and other drug use, and smoking. The prevalence of physician substance use has been of interest not only for the impact that it has on the individual engaging in the behavior, but also because of the consequences it can have on the provision of care. The issue of impairment of medical faculty is given in more detail in Chapter 3, the following is a summary of the substance use behavior as reported in the literature.

The reported substance use among medical faculty varies widely. This may be due to the sources of these data. Estimates of substance abuse are often derived from cause-specific mortality and morbidity reports (e.g., reports of cirrhosis of the liver), health care utilization (admissions to drug and alcohol treatment programs), and surveys of self-reports of alcohol/drug consumption. Each one of these, at best, offers incomplete pictures of the experience of alcohol and drug use. Finding an appropriate comparison group to assess relative frequency of substance use/abuse is also challenging. Brewster (1986) after conducting an exhaustive literature review on substance use in the physician population conclude that survey data may be in fact the best way to assess the prevalence of these behaviors [30].

Hughes et al. conducted an anonymous mail survey with a national sample of physicians selected from the AMA master file to estimate the prevalence of substance use [31]. Alcohol was the most frequently used substance (87% had consumed alcohol in the past year). Self-reported lifetime alcohol *abuse* was reported by 6% of the sample, and in past year less than 2%. This is significantly less than the estimates of alcohol abuse in the general population at the time (13–16%). However, 10% reported daily alcohol use, and 9% indicated having five or more drinks per day at least once in the past month. Minor opiates were the most frequently used prescription drugs; 8% had used minor opiates without the supervision of another physician in the past year, 4.5% in the past month. The authors suggest the pattern and reasons given for prescription drug use point to self-treatment.

Several smaller studies also report a prevalence of alcohol/substance use problems lower than that found in the general public, but there are reports of heavy alcohol use and elevated use of prescription drugs without another physician's supervision. McAuliffe reported on a survey given to a random sample of physicians from a state medical society membership roster and medical students in the same state [32]. A comparison group of pharmacists randomly selected from the state's pharmaceutical society membership list and pharmacy students in area schools was also surveyed. Respondents were asked about current (within the past year) and lifetime drinking practices, amount consumed, and qualitative self-assessment of their behaviors regarding abuse and dysfunction. Physicians and medical students reported drinking more frequently, but consumed less per drinking episode. Less than 3% of any of the groups reported having a current drinking problem. Practitioners drank more often than students, but less quantity per episode. There were more abstainers among pharmacists (4% of doctors, 5% of medical students, 10% of pharmacists, and 11% of pharmacy students).

Reinhardt (2005) in the study mentioned above of faculty and residents at one medical center in California, found that 6% had a high likelihood of harmful alcohol consumption using the AUDIT scale. Almost 5% acknowledged the use of sedatives or hypnotics without a prescription in the last 12 months [17].

Data from the WPHS reveal that women physicians drank alcohol more frequently than both high SES and nonhigh SES women (on an average 8.5, 7.2, and 6.1 days per month, respectively), but at lesser quantities [29]. Abuissa and colleagues' study of cardiologists in a large coalition of cardiology groups reported a high prevalence of regular alcohol use (72% consumed greater than or equal to one drink/week) and 50% drank daily [33]. Schindler found that older faculty had increased alcohol consumption compared to younger faculty (24% under age 45, drinking daily to several times of week compared to 45% of those over 55; 36% of males drank daily or several times a week compared to 27% of women) [15].

The reports of substance abuse or problems are low compared to that of the general public. However, the consistent findings that physicians are using alcohol more frequently than the general public or their counterparts in other professions need further investigation.

Smoking practices have been of great interest in the general population and among physicians as well. In general, physicians have lower rates of smoking than the general public and smoking has decreased among the physician population faster than in other health professions. In an analysis of National Health Interview Survey (NHIS) data, Nelson et al. reviewed trends of smoking from 1974 to 1991 among physicians, registered nurses, and practical nurses [34]. Smoking prevalence among physicians declined from 19% in the late 1970s to 3% in 1991, an average decline of 1.5% per year. For registered nurses in the same time frame the decline was from 31% to 18%. Estimates of current smoking in the general US population from the NHIS 2006 survey indicate 21% of the general US adults and 6.6% of respondents with a graduate degree [35]. The authors suggest that this decline in the smoking prevalence among health professionals is due to the increase both in the number of health professionals quitting smoking and in the number of nonsmokers entering the heath care workforce. Other studies of US physicians also report the prevalence of current smokers under 5% for both genders. In the WPHS, women physicians were less likely to smoke than both women from high SES and all women in the BRFSS; 3.7% of physicians vs. 8% of high SES women, and 25% of women not of high SES were current smokers [29].

Implications

The experiences of stress and the practice of poor health behaviors can have devastating effects on one's health and well-being. In addition, in the case of academic medical faculty the impact can reach beyond that of the individual's health to affect the provision of care to the patient, relationships with colleagues and students with whom they interact on a daily basis. The effect on the provision of care can manifest itself in different ways. A study of internal medicine residents showed that physicians who experience burnout were more likely to self-report patient care practices and attitudes that suggested suboptimal care [36]. While physicians that engage in preventive health behaviors are more likely to encourage the same among their patients. Frank (2000) found that the most powerful predictor of a physician engaging in preventive health behavior counseling with his/her patients is in fact the physicians own preventive health practices [37]. This relationship has been shown to be true with counseling regarding exercise, seat-belt use, and smoking [38, 39].

Conclusions

In general, physicians enjoy longer lives and healthier lifestyles than their counterparts in the general population. However, several concerns arise from this review and warrant further analysis and intervention. First, as in the general population, persistent racial disparities exist in the experience of morbidity and mortality among physicians (see Chapter 8, for further discussion). Second, where academic medical faculty experience elevated rates of morbidity and mortality they are of a preventable nature, i.e., suicide and accidents. Finally, the reluctance to seek care when needed, particularly mental health services, and the reliance on self-care are troublesome.

Further research is needed on the prevalence and incidence of morbidities and cause-specific mortality in academic medicine faculty, specifically the nonphysician, female, and minority faculty members. Additional studies identifying barriers to seeking care and elucidation of both risk and protective factors are also needed. Finally, prospective studies on the experience of wellness and clarification of health practices and lifestyle choices among academic medical faculty could inform effective interventions.

We turn again to the questions with which we began this chapter, "Who are medical faculty? And how do we define their health?" Medical faculty are a diverse group of individuals working in an ever-changing and evolving environment. The elevated experiences of stress in dealing with this environment have an impact on faculty health. A comprehensive approach that addresses all aspects of health and well-being for this diverse population is needed to ensure not only the health of the faculty members themselves, but also the well-being of the students they mentor and the patients they serve. Only a "healthy faculty" will be able to successfully meet the clinical, research, and teaching missions of the academic medical center.

References

1. AAMC (2007) US Medical School Faculty Roster Report 2006 http://www.aamc.org/data/facultyroster/usmsf07/usmsf07.htm

2. Ostun, T.B. (2001) What is health? how do we measure health? *Sozial- Und Praventivmedizin* **46(2)**, 71–72.
3. Frank, E., Biola, H., and Burnett, C.A. (2000) Mortality rates and causes among US physicians *American Journal of Preventive Medicine* **19(3)**, 155–159.
4. Calvert, G.M., Merling, J.W., and Burnett, C.A. (1999) Ischemic heart disease mortality and occupation among 16- to 60-year-old males *Journal of Occupational & Environmental Medicine* **41(11)**, 960–966.
5. Williams, S.V., Munford, R.S., Colton, T., Murphy, D.A., and Poskanzer, D.C. (1971) Mortality among physicians: a cohort study *Journal of Chronic Diseases* **24**, 393–401.
6. Ullmann, D., Phillips, R.L., and Beeson, L., et al. (1991) Cause-specific mortality among physicians with differing life-styles *Journal of the American Medical Association* **265**, 2352–2359.
7. Sankoff, J.S., Hockenberry, S., Simon, L.J., and Jones, R.L. (1995) Mortality of young physicians in the United States, 1980–1988 *Academic Medicine* **70**, 242–244.
8. Lindeman, S., Laara, E., Hakko, H., and Lonnqvist, J. (1996) A systematic review on gender-specific suicide mortality in medical doctors *British Journal of Psychiatry* **168(3)**, 274–279.
9. Center, C., Davis, M., Detre, T., Ford, D.E., Hansbrough, W., and Hendin, H., et al. (2003) Confronting depression and suicide in physicians: A consensus statement *Journal of the American Medical Association* **289(23)**, 3161–3166.
10. Schernhammer, E.S., and Colditz, G.A. (2004) Suicide rates among physicians: a quantitative and gender assessment (meta-analysis) *American Journal of Psychiatry* **161(12)**, 2295–2302.
11. Vishnevskaia, E.P., and Atiakina, I.K. (1972) Data on the morbidity of the professor-instructor staff of the I. M. Sechenov First Moscow Medical Institute (Russian). *Gigiena i Sanitariia* **37(7)**, 100–102.
12. Linn, L.S., Yager, J., Cope, D., and Leake, B. (1985) Health status, job satisfaction, job stress, and life satisfaction among academic and clinical faculty *Journal of the American Medical Association* **254(19)**, 2775–2782.
13. Aasland, O.G., Olff, M., Falkum, E., Schweder, T., and Ursin, H. (1997) Health complaints and job stress in Norwegian physicians: the use of an overlapping questionnaire design *Social Science & Medicine* **45(11)**, 1615–1629.
14. Sutherland, V.J., and Cooper, C.L. (1992) Job stress, satisfaction, and mental health among general practitioners before and after introduction of new contract *BMJ* **304**, 1545–1548.
15. Schindler, B.A., Novack, D.H., Cohen, D.G., Yager, J., Wang, D., and Shaheen, N.J., et al. (2006) The impact of the changing health care environment on the health and well-being of faculty at four medical schools *Academic Medicine* **81(1)**, 27–34.
16. Blazer, D.G., Kessler, R.C., McGonagle, K.A., and Swartz, M.S. (1994) The prevalence and distribution of major depression in a national community sample: the National Comorbidity Survey *American Journal of Psychiatry* **151**, 979–986.
17. Reinhart, T., Chavez, E., Jackson, M., and Mathews, W.C. (2005) Survey of physician well-being and health behaviors at an academic medical center *Med Educ Online* (serial online) **10(6)**, http://www.med-ed-online.org
18. Ford, D.E., Mead, L.A., Chang, P.P., Cooper-Patrick, L., Wang, N.Y., and Klag, M.J. (1998) Depression is a risk factor for coronary artery disease in men: the precursors study *Archives of Internal Medicine* **158**, 1422–1426.
19. Williams, E.S., Konrad, T.R., Linzer, M., McMurray, J., Pathman, D.E., Gerrity, M., Schwartz, M.D., Scheckler, W.E., and Douglas, J. (2002) Physician, practice, and patient characteristics related to primary care physician physical and mental health: Results from the physician worklife survey *Health Services Research* **37(1)**, 121–43.
20. Lazarus, R.S., and Folkman, S. (1984) *Stress, Appraisal, and Coping* New York: Springer.
21. Ivancevich, J.M., and Matteson, M.T. (1980) *Stress and Work: A Managerial Perspective* Glenview: Scott Foresman.
22. Kahn, K.L., Goldberg, R.J., DeCosimo, D., and Dalen, J.E. (1988) Health maintenance activities of physicians and nonphysicians *Archives of Internal Medicine* **148(11)**, 2433–2436.

23. Gross, C.P., Mead, L.A., and Ford, D.E., et al. (2000) Physician, heal thyself? Regular source of care and use of preventive health services among physicians *Archives of Internal Medicine* **160**, 3209–3214.
24. Wachtel, T.J., Wilcox, V.L., Moulton, A.W., Tammaro, D., and Stein, M.D. (1995) Physicians' utilization of health care *Journal of General Internal Medicine* **10(5)**, 261–265.
25. Wang, P.S., Lane, M., Olfson, M., Pincus, H.A., Wells, K.B., and Kessler, R.C. (2005) Twelve-month use of mental health services in the United States: Results from the national comorbidity survey replication *Archives of General Psychiatry* **62(6)**, 629–640.
26. Givens, J.L., and Tjia, J. (2002) Depressed medical students' use of mental health services and barriers to use *Academic Medicine* **77(9)**, 918–921.
27. Christie, J.D., Rosen, I.M., and Bellini, L.M., et al. (1998) Prescription drug use and self-prescription among resident physicians *Journal of the American Medical Association* **280**, 1253–1255.
28. Steering Committee of the Physicians' Health Study Research Group (1989) Final report on the aspirin component of the ongoing Physicians' Health Study *New England Journal of Medicine* **321**, 129–135.
29. Frank, E., Brogan, D.J., Mokdad, A.H., Simoes, E.J., Kahn, H.S., and Greenberg, R.S. (1998) Health-related behaviors of women physicians vs other women in the United States *Archives of Internal Medicine* **158**, 342–348.
30. Brewster, J.M. (1986) Prevalence of alcohol and other drug problems among physicians *Journal of the American Medical Association* **255**, 1913–1920.
31. Hughes, P.H., Brandenberg, N., and Baldwin, Jr., D.C., et al. (1992) Prevalence of substance use among US physicians *Journal of the American Medical Association* **267**, 2333–2339.
32. McAuliffe, W.E., Rohman, M., Breer, P., Wyshak, G., Santangelo, S., and Magnuson, E. (1991) Alcohol use and abuse in random samples of physicians and medical students *American Journal of Public Health* **81(2)**, 177–182.
33. Abuissa, H., Lavie, C., Spertus, J., and O'Keefe Jr., J. (2006) Personal health habits of American cardiologists *American Journal of Cardiology* **97(7)**, 1093–1096.
34. Nelson, D.E., Giovino, G.A., Emont, S.L., Brackbill, R., Cameron, L.L., Peddicord, J., and Mowery, P.D. (1994) Trends in cigarette smoking among US physicians and nurses *Journal of the American Medical Association* **271(16)**, 1273–1275.
35. Centers for Disease Control (2007) Cigarette Smoking Among Adults – United States, 2006 *Morbidity and Mortality Weekly Report* **56(44)**, 1157–1161.
36. Shanafelt, T.D., Bradley, K.A., Wipf, J.E., and Back, A.L. (2002) Burnout and self-reported patient care in an internal medicine residency program *Annals of Internal Medicine* **136(5)**, 358–367.
37. Frank, E., Rothenberg, R., Lewis, C., and Belodoff, B.F. (2000) Correlates of physicians' prevention-related practices *Archives of Family Medicine* **9**, 359–367.
38. Lewis, C.E., Clancy, C., Leake, B., and Schwartz, J.S. (1991) The counseling practices of internists *Annals of Internal Medicine* **114(1)**, 54–58.
39. Wells, K.B., Lewis, C.E., Leake, B., and Ware Jr., J.E. (1984) Do physicians preach what they practice? A study of physicians' health habits and counseling practices *Journal of the American Medical Association* **252**, 2846–2848.

Chapter 3
Causes and Treatment of Impairment and Burnout in Physicians: The Epidemic Within

Eugene V. Boisaubin

Abstract No one knows the exact numbers or percent of physicians who are impaired in academic medicine. There is no reason to believe that it is significantly different from the national figures—15% throughout a professional career. Although the scope of "impairment" is growing wider, this chapter will focus on the traditional categories of drug and alcohol abuse, mental illness, particularly depression and suicide, and the newer category of the "disruptive physician." This chapter will also link predisposing issues such as stress and burnout to the development of impairment and identify issues unique to the academic world that might foster the development of these impairments. These might include a very powerful drive for achievement, exceptional conscientiousness, an ability to deny personal problems, and an inability to achieve a balance between the professional and personal life. Identifying and even treating these problems before they evolve into true impairment is a continual challenge. Society and the professions are raising the standards of what they expect from their professionals, including physicians, and these include accountability, transparency, and expectations of high levels of function without impairment that could affect patient care. There is also an obligation to help our fellow professionals, particularly in academic settings, to be able to fulfill their lives with both professional and personal satisfaction.

Keywords Impairment, substance abuse, disruptive physician, predisposing issues, burnout

Introduction

Historically, physicians have struggled to find an adequate balance between their professional and personal lives. Although it is virtually impossible to compare adequately the stresses and demands upon physicians of previous generations with those

E.V. Boisaubin
Professor of Internal Medicine, Department of Internal Medicine, University of Texas-Houston Medical School, Houston, Texas, USA
e-mail: Eugene.Boisaubin@uth.tmc.edu

T.R. Cole et al. (eds.) *Faculty Health in Academic Medicine*,
© Humana Press, a part of Springer Science + Business Media, LLC 2009

of the current world, many indicators suggest that at least the variety orspectrum of stressors has increased significantly in the past 20 years [1]. Professional pressures include practicing in an increasingly competitive, fast-paced, cost-efficient manner, surrounded by a sea of regulatory restraints, often based on the insurance companies' policies. In addition, there are the expectations of the institution, patient, and family. Given these stresses, those practices or behaviors that might compromise physicians' skills and compassion should be identified, addressed, and corrected if possible.

The personal life of the physician is not necessarily a supportive or energizing means for balance. Direct and indirect pressures upon American family life appear to be increasing and include not only the workplace demands, but also expectations upon children to engage in their multitude of activities. As well, there has to be time for friends, church, and community. How is it possible for physicians these days to carry out their professional role well and at the same time carry out a successful personal life, all the while remaining free of unnecessary stresses or even impairment?

Impairment

Physician impairment results when a set of attitudes and behaviors, personal and environmental, usually caused or aggravated by a disease, harms physicians, negatively affects their families, and even compromises patient care. Typical causes of impairment include drug and alcohol abuse and/or mental illness. Impairment that is early in its evolution might be readily treatable without a demonstrable impact upon the physician's personal welfare, family well-being, and patient care. In contrast, if the impairment is severe and is not identified and treated in a timely fashion, the physician may experience loss of licensure and career, professional censure, loss of family structure and personal identity, and possibly even death by suicide, drug overdose, or secondary health consequences and other complications of the underlying disease [2].

Professional organizations such as the American Medical Association (AMA) define the impaired physician as one who is "unable to fulfill professional or personal responsibilities because of a psychiatric illness, alcoholism, or drug dependency [3]." This definition focuses only on substance abuse and mental illness, although virtually any significant medical situation or even chronic, unremitting stress-related problems can affect judgment and performance, thereby compromising the capability to provide optimal medical care. However, here the focus will be on substance abuse, depression, stress, burnout, and sleep deprivation, since these conditions often are combined and interactive [4]. Impairment in academic medicine, including both faculty and resident physicians, will be primarily addressed although more data exists for the totality of clinical practice.

Magnitude of the Problem

The precise number of impaired physicians in America, or any locality, is unknown and is hard to document precisely for several reasons. Some physicians who have come forward to seek help on their own have entered treatment confidentially and have avoided becoming a statistic for any state or licensing group. Second, the above definition is neither precise nor inclusive. Becoming impaired is not an overnight event, but rather a slow process that may continue to evolve for months or years and sometimes decades before the full impact is realized. Impairment is also not an all or none phenomenon. A high level of alcohol consumption, for example, may continue unchanged for years with no obvious deleterious effects upon personal or professional life. But at some point consumption increases or a complication occurs and suddenly others may now be made aware of what they perceive as a new problem.

Also, in part, because of the imprecision of the definition, a number of impaired physicians may not be appropriately identified as having a problem and may simply leave medicine on their own, or might be forced to retire because of associated medical conditions, never having a formal diagnosis. Prevalence of physician impairment involving drugs or alcohol reported in the published literature is also imprecise. Figures range from lows of 2% to highs of 18%. A large national survey of 9,600 physicians indicated that 2% reported substance abuse or dependence problems in the past year and 8% at some time in their life. However, 9.3% reported having five or more drinks a day at least once in the past month [5]. Rates as high as 12.9% come from self-reports of 1,014 male medical student graduates of Johns Hopkins University at age 52–68 [6]. In California, a state review group found a peak rate of 18% [7]. These higher figures are more consistent with the 13.5% rate of alcohol disorders in the adult population reported from the National Institute of Mental Health Study [8]. Since alcoholism is increasingly viewed as a medical disorder with genetics playing a powerful role, the similar rates between physicians and the general population should not be surprising.

Prevalence of Depression and Suicide

Although the data are again imprecise, most surveys suggest that the rate of disorders related to anxiety or depression in American physicians are at least equal to the national rate of approximately 15–20% [9]. One retrospective study of physician admissions to psychiatric hospitals and clinics suggests that doctors have higher rates of depression compared to the general population [10]. A landmark prospective study followed Harvard sophomores for over 30 years. Forty-seven became physicians, and when compared to their peers, they showed higher rates of poor marriage, drug and alcohol abuse, and the use of psychotherapy [11]. In almost all studies of this kind, the rates of depressive and/or anxiety disorders is approximately twice as common among women in medicine as men, a finding which parallels data for the general population in America [12, 13] As the presence of women in medicine continues to increase annually, growing numbers of women are appearing for diagnosis and treatment. Less data are available for women concerning the long-term impact of these illnesses, treatment, and recovery since a relatively small number of women were present in the early studies. Studies of the mental health of medical students and house officers have also revealed high rates of depression during training, particularly during the most stressful periods [14, 15]. Some depressive and anxiety symptoms are virtually universal among individuals in medical training, with the majority of these being only temporary, for example, during difficult clinical rotations [16, 17]. For those who have more sustained signs and symptoms, we assume that their biologic predispositions as well as certain continuing stress factors would be contributory.

Suicide is the most tragic outcome for the impaired physician. Several reviews have reported that suicide rates among physicians were higher than those in the general population, and higher than those of other professionals [18, 19]. Most striking is the high rate in female physicians, found in one study to be three to four times that of women in the general population [20]. There is an increasing concern that the elevating levels of stress, frustration, and burnout in contemporary academic medicine, medical training, and practice may further aggravate the above problem. However, hard and stressful work alone does not result in suicide. Those who do commit suicide almost always have significant identifiable underlying mental illnesses, such as major depression and/or bipolar disorders, usually coupled with alcoholism and major drug abuse.

Burnout Among Physicians

For over 25 years, articles have appeared, at first periodically, and now more regularly in the medical literature, concerning the impact of stress on the professional and personal lives of physicians. Stress of course is universal in human existence. Recently, interest has turned to a more focused aspect of continued unrelenting stress that being the phenomenon of "burnout". At first only a concept, it has evolved over the last decade into a valid psychological diagnosis, replete with validation studies concerning its accuracy. The greatest work has been done by Maslach who created the Maslach Burnout Inventory Manual, which has been used since 1996 as a standard for measuring this condition [21]. In essence, burnout is a syndrome defined by the three principal components: emotional exhaustion, depersonalization, and diminished feelings of personal accomplishment. At variance with major depressive disorders, which impact all aspects of a patient's life, burnout is a work-related syndrome. It is also more likely to occur in jobs that require extensive care of other people. Burnout is diagnosed on the Inventory by the combination of high scores for emotional exhaustion and depersonalization, and low scores for personal accomplishment. Since the late 1980s, at least 30 articles in major journals have addressed the issue of burnout in physicians, but primarily residents in training.

The combined evidence indicates that the components of burnout are common among physicians. Surveys of practicing physicians indicate that 46–80% report moderate to high levels of emotional exhaustion, 22–93% report moderate to high levels of depersonalization, and 16–79% report low to moderate levels of personal achievements [22–24]. Studies of academic physicians have yielded similar results. In a survey of 119 academic obstetrics and gynecology department chairs in the United States and Puerto Rico, with a response rate of 91%, Gabbe et al. found that 56% of respondents demonstrated high levels of emotional exhaustion, 36% had high levels of depersonalization, and 21% reported low levels of personal accomplishment [25]. Studies of department chairs in pediatrics and otolaryngology have also identified significant levels of stress and burnout [26, 27]. In a longitudinal study of academic administrators, Mirvis et al. reported an increase in the prevalence of burnout from 25.3% in 1989 to 38.1% in 1997 from a cohort of 83 administrators in Department of Veteran's Affairs Medical Centers [28]. Unfortunately, virtually no studies can be identified that look specifically at these issues in other basic science or non-clinical faculty in academic medicine, although we know the administrative and financial pressures are similar.

Articles that specifically looked at academic centers have most commonly focused upon residents in training. In 2002, Shanafelt et al. reported on the rate of burnout and self-reported patient care in an internal medicine residency program at the University of Washington in Seattle [29]. The Maslach survey was sent to all 151 residents with a 76% response rate, and 76% also met criteria for burnout. Burned out residents were also more likely to self-report providing suboptimal patient care and 50% of these residents also had depressive symptoms, and 9% described at risk alcohol use, and more career dissatisfaction. The three greatest stresses in training were identified as inadequate sleep (41%), shifts greater than 24 hours (40%), and inadequate leisure time (42%). Residents who meet burnout criteria were significantly more likely to rate these stresses as important. Shanafelt also reported that 51% and 31% of burned out residents had a positive result on a depression screen and self-reported major depression, respectively, versus 29% and 11% of residents who were not burned out. Burned out residents were also significantly more likely to indicate that they had been responsible for one event of suboptimal patient care at least weekly or monthly compared with non-burned out residents. Whether suboptimal care was really provided, or whether it was only the resident's perception, is not clear. Additional factors uncovered when surveying internal medicine program directors include a burgeoning financial burden, with resident debt commonly exceeding $50,000 and moonlighting becoming a common compensatory mechanism [30].

The specific risk factors for physician burnout have not been clearly established, but suggestive factors are present. In their longitudinal study, Mirvis et al. found that younger age, lack of role clarity, and perceived inadequacy of resources all predicted the development and progression of burnout [29]. The two studies quoted above of chairmen in pediatrics and otolaryngology noted similar patterns as well as self-reported work load. Those results however may reflect "survival bias" where those who burn out early in their careers are more likely to quit their jobs, then selecting for older respondents with lower levels of burnout. Finally, Gabbe et al.

found that high emotional exhaustion in obstetrics and gynecology department chairs was inversely related to the self-reported level of spousal support [25]. Walter Baile, M.D., a professor of psychiatry at M. D. Anderson Cancer Center in Houston and others have described stress factors for oncology faculty. These traits include working very long hours, denying their own personal needs, finding it difficult to ask for help, sacrificing their family and personal life for their jobs, taking a tremendous amount of responsibility for their patients, being very self-critical, and being unable to say "no" to requests for additional time and effort [31].

Sleep Deprivation

Sleep deprivation has been identified as perhaps the most important physiologic component of stress and burnout. Increasingly active lives with demands for time and commitment have stretched human biologic clocks to the point that the great majority of Americans receive inadequate sleep. This situation is particularly true for health professionals in training, such as residents. The sleep debt, derived from abnormal sleep quality as well as quantity, can accumulate over a number of days and weeks, thereby producing a chronic type of impairment. "On call" schedules, with up to 36 hours of sleep deprivation, are particularly detrimental, including scheduled periods of acute sleep deprivation, chronic partial sleep loss, cumulative sleep debt, misalignment of circadian rhythms, and medical performance required within minutes of awakening [32]. Residents may be equivalent in sleep deprivation to patients with serious sleep disorders such as sleep apnea and narcolepsy. Sleep deprivation in residents, faculty, and practitioners has been associated with an accelerated decline in clinical performance and may even affect patient care [33, 34].

Treatment Options: One University's Response

The treatment options for physicians' problems such as substance abuse, mental illness, burnout, and even sleep deprivation are far more varied and available today than in the past. In a subsequent chapter in this volume, Goodrich outlines programs and interventions that are currently available throughout the country to help physicians in their attempt to restore energy, meaning, fulfillment, and purpose in their professional careers and personal lives. The remainder of this chapter will address other options, such as approaches to substance abuse and mood disorders, and describe how one university and specifically one department, have worked to solve problems involving resident and faculty stress, burnout, and impairment.

Diagnosing substance abuse requires careful corroboration of behavior, a clear diagnosis of any underlying medical conditions, such as major depression coexistent with drug abuse, and a highly organized treatment program [9]. Physicians with documented problems, whether in private practice or academia, currently must be entered into effective treatment programs with regular monitoring that can be reported back to supervisors and/or department chairs. The standard treatment programs now include an initial inpatient evaluation, if necessary, or more routinely, in an outpatient setting, continuing with regular outpatient follow-up with an addictionologist or psychiatrist if mental illness is also involved. Routine drug testing is now mandatory in all programs where substances are involved with a carefully identified reporting schema. Contracts are created between the individual and the supervising institution, whether a hospital or clinical department, with the expectations for treatment, follow-up, and recurrence being well-outlined. Recurrence is almost inevitable with a chronic relapsing disease such as addiction. The number of additional opportunities the physician might receive after a relapse for re-entering treatment need to be specified in writing.

Important mental illness, such as major depression and/or bipolar disorders, need to be evaluated and treated by a psychiatrist. Clear lines concerning follow-up, the creation of a contract, and regular progress reports to individuals in a supervisory capacity are necessary. Again, contingencies for relapse need to be anticipated.

At the University of Texas Medical School at Houston, a formal agreement between the resident and fellow population and the University's Employee Assistance Program (EAP) has been in effect now for over four years, focusing on impairment and well-being. In the Department of Medicine alone, 13 residents have been referred to the EAP in the last three years. Although any resident at any time may contact the EAP for a variety of services, including financial planning, personal relationship issues, or child care, there are three formal kinds of referral through the Department. First, if the resident is noted to either portray or describe stress in his or her professional or personal life, a referral to the EAP is recommended, but not required. If the issue is more significant, such as problems with behavior and attitude, including anger, the physician can be referred through the Department to the EAP for an evaluation with a required summary report returning to the program directors. The third and most serious category involves referral for very significant problems such as identified substance abuse, extreme behavior problems, or repetitive signs and symptoms of real or possible mental illness. In this situation not only is an evaluation required, but an ongoing care plan has to be constituted including regular use of drug testing if necessary and regular counseling reports that go directly to the program director. The ability of the trainee to continue in and complete the academic program is dependent upon successful adherence to these requirements, also formalized with a contract.

The most common reason for referral in the past four years has been behavior problems, particularly anger management, afflicting six of the referred residents. These referrals usually result in initial counseling with subsequent involvement of a psychiatrist if a major medical illness is identified. Alternatively, a local program that specializes in anger management may be utilized. Proof of completion of the program is required for the physician to continue in training. Substance abuse, depression, and adjustment reactions are the primary reasons for the remaining referrals. One resident has been terminated from the residency program in the past three years because of refractory behavior problems and an unwillingness to abide by treatment plans. The same programs described above are available for the Departmental faculty, although they almost always prefer referral to a private counselor or program because of confidentiality. Increasingly important are factors that discourage faculty evaluation and treatment, such as new awareness of the problem by malpractice insurance companies, a hospital credentialing committee, or even the state medical board. Enthusiastic support and commitment for assistance programs for faculty must come from senior administration such as departmental chairs and the dean's office.

The Department has also created a series of preventive educational steps to help young physicians in training identify early symptoms or problems with the typical disorders noted above and to seek advice and/or counseling at an early point. Educational steps include two formal presentations by faculty during orientation on stress and burnout, and the availability of assistance and treatment facilities, particularly the EAP. During the presentations on ethics and professionalism given to all 150 members of the medicine house staff residency, continued references are made to the importance of the psychological state of the physician including the attitudes and behaviors that often determine professional behavior.

Twice a year, as required by the Residency Review Committee (RRC), which accredits national training programs, each resident meets with one of the program directors and specific questions are now asked concerning stress and burnout, and how both the personal and professional lives are progressing. Other sources of information concerning resident attitudes and behaviors include clinical student evaluations of the residents, monthly faculty evaluations, and an annual anonymous peer review survey.

This past year saw the initiation of a trial project, through the McGovern Center for Health, Humanities and the Human Spirit, to introduce a sacred vocation program for residents. Small group sessions were carried out with a sample of medicine residents to encourage them to explore the meaning and purpose of their work and how their work is related to their personal lives. Responses from the residents concerning the experiment were extremely positive. Although this program is directed toward residents only, faculty self-help groups are increasing in popularity and are discussed elsewhere in the book.

Last, awareness of sleep deprivation, particularly in residents in training, is being made a priority at the national level. There is a formal educational session during orientation for all residents beginning the training program at this University. There are also now strict supervisory guidelines, instituted by the RRC for all residency programs across the country, involving the limitation of all house staff work to fewer than 80 hours a week averaged over a month, as well as limits to consecutive call hours and the required one day a week off from clinical responsibilities [32, 33]. Early feedback suggests that residents are more satisfied with these hour restrictions although violations do occur and must be followed up by the program directors for continued accreditation.

A major challenge for program directors and hospital administration is the difficulty now evolving in continuity of patient care caused by restrictions on residents' work hours. In essence, the care of patients may be increasingly divided and partitioned between serial physicians or even groups of physicians rather than one care provider for a longer period of time. Although the series of providers is undoubtedly better rested and probably able to make better clinical decisions, oversights in patient care are not uncommon with a serial multi-provider model when the patient is periodically "handed off" to new care givers. Training programs in America will have to continue to search for the optimal way to balance the need for adequate rest time for residents with high-quality patient care. Unfortunately, there is no parallel limitation on work hours or reporting system for academic faculty who, as noted, may also suffer the effects of long hours and sleep deprivation. Ultimately, although there are important differences between physicians in training and academic practice, many of the factors that lead to stress and impairment are shared by both groups [35]. It is hoped that programs on improved well-being can be shared as well.

References

1. Murray, A., Montgomery, J.E., and Chang, H. (2001) Doctor discontent: a comparison of physician satisfaction in different delivery system settings: 1986 and 1997 *Journal of General Internal Medicine* **16**, 452–459.
2. Coombs, R.H. (1997) *Drug Impaired Professionals* Cambridge: Harvard University Press, 3–30.
3. American Medical Association Council on Mental Health (1973) The sick physician: impairment by psychiatric disorders including alcoholism and drug dependency *Journal of the American Medical Association* **223**, 684–687.
4. Regier, D.A., Farmer, M.E., Rae, D.S., et al. (1990) Comorbidity of mental disorders with alcohol and other drug abuse: results from the Epidemiologic Catchment Area study *Journal of the American Medical Association* **264**, 2511–2518.
5. Hughes, P.H., Brandenberg, N., Baldwin, D.C., et al. (1992) Prevalence of substance use among US physicians *Journal of the American Medical Association* **267**, 2333–2339.
6. Moore, R.D., Mead, L., and Pearson, T.A. (1990) Youthful precursors of alcohol abuse in physicians *American Journal of Medicine* **88**, 332–336.
7. Medical Board of California (1995) *The Medical Board's Diversion Program. Mission statement* Sacramento.
8. McAuliffe, W.E., Rohman, M., Breer, P., Wyshak, G., Santangelo, S., and Magnuson, E. (1991) Alcohol use and abuse in random samples of physicians and medical students *American Journal of Public Health* **81**, 177–182.
9. Boisaubin, E.V., and Levine, R.E. (2001) Identifying and assisting the impaired physician *American Journal of the Medical Sciences* **2**, 31–36.
10. Murray, R.M. (1977) Psychiatric illness in male doctors and controls; an analysis of Scottish hospital inpatient data *British Journal of Psychiatry* **131**, 1–10.
11. Vaillant, G.E., Sobowale, N.C., and McArthur, C. (1972) Some psychologic vulnerability of physicians *New England Journal of Medicine* **287(8)**, 372–375.
12. Clayton, P.J., Marten, S., Davis, M.A., and Wochik, E. Mood disorders in women professionals *Journal of Affective Disorders* **2**, 37–47.
13. Kessler, R.C., McGonagle, K.A., Zhao, S., Nelson, C.B., Hughes, P.H., Eshleman, M.A., Wittchen, H., and Kendler, K.S. (1994) Lifetime and 12 month prevalence of DSM-III-R psy-

chiatric disorders in the United States; results from the National Comorbidity Survey *Archives of General Psychiatry* **51(1)**, 8–19.

14. Valko, R.J., and Clayton, P.J. (1975) Depression in internship *Diseases of the Nervous System* **36**, 26–29.
15. Reuben, D.B. (1985) Depressive symptoms in medical house officers *Archives of Internal Medicine* **145**, 286–288.
16. Schneider, S.E., and Phillips, W.M. (1993) Depression and anxiety in medical, surgical, and pediatric interns *Psychological Reports* **72**, 1145–1146.
17. Hendrie, H.C., Clair, D.K., Brittain, H.M., and Fadul PE. (1990) A study of anxiety/depressive symptoms of medical students, house staff, and their spouses/partners *Journal of Nervous and Mental Disease* **178(3)**, 204–207.
18. Lindeman, S., Laara, E., Hakko, H., and Lonnqvist, J. (1996) A systematic review on gender-specific suicide mortality in medical doctors *British Journal of Psychiatry* **168**, 274–279.
19. Steppacher, R.C., and Mausner. J.S. (1974) Suicide in male and female physicians *Journal of the American Medical Association* **228(3)**, 323–327.
20. Schernhammer, E. (2005) Taking their own lives-the high rate of physician suicide *New England Journal of Medicine* **352(24)**, 2473–2476.
21. Maslach. C., Jackson, S.E., and Leiter, M.P. *Maslach Burnout Inventory Manual* 3rd ed. Palo Alto, CA: Consulting Psychologies Press.
22. Linzer, M., Konrad, T.R., Douglas, J., et al. (2000) Managed care, time pressure and physician time satisfaction *Journal of General Internal Medicine* **15**, 441–450.
23. Campbell, D.A., Sonnad, S.S., Eckhauser, F.E., et al. (2001) Burnout among American surgeons *Surgery* **130**, 696–705.
24. Deckard, G.J., Hicks, L.L., and Hamory, B.H. (1992) The occurrence and distribution of burnout among infectious disease physicians *Journal of Infectious Diseases* **165**, 224–228.
25. Gabbe, S.G., Melville, J., Mandel, L., and Walker, Z. (2002) Burnout in chairs of obstetrics and gynecology: diagnosis, treatment and prevention *American Journal of Obstetrics and Gynecology* **186(4)**, 601–12
26. McPhillips, H.A., Stanton, B., Zuckerman, B., and Stapleton, F.B. (2007) Role of a pediatric department chair: factors leading to satisfaction and burnout *Journal of Pediatrics* **151(4)**, 425–430.
27. Johns, M.M., and Ossott, R.H. (2005) Burnout in academic chairs of otolaryngology *Laryngoscope* **115(11)**, 2056–2061.
28. Mirvis, D.M., Graney, M.J., and Kilpatrick, A.O. (1999) Burnout among leaders of the Department of Veterans Affairs medical centers; contributing factors as determined by a longitudinal study *Journal of the Health and Human Services Administration* **21**, 390–412.
29. Shanafelt, T.D., Bradley, K.A., Wipf, J.E., and Back, A.C. (2002) Burnout and self-reported patient care in internal medicine residency programs *Annals of Internal Medicine* **136**, 358–367.
30. Resident Services Committee, Association of Program Directors in Internal Medicine (1988) Stress and impairment during residency training: strategies for reduction, identification and management *Annals of Internal Medicine* **109**, 154–161.
31. Kuerer, H.M., Eberlein, T.J., Pollock, R.E., Huschka, M., Baile, W.F., Morrow, M., Michelassi, F., Singletary, S.E., Novotny, P., Sloan, J., and Shanafelt, T.D. (2007) Career satisfaction, practice patterns and burnout among surgical oncologists: report on the quality of life of members of the Society of Surgical Oncology *Annals of Surgical Oncology* **14(11)**, 3043–3053.
32. Papp, K.K., Stoller, E.P., Sage, P., Aikens, J.E., Owens, J., Avidan, A., Phillips, B., Rosen, R., and Strohl, K.P. (2004) The effects of sleep loss and fatigue on resident-physicians: a multi-institutional, mixed method study *Academic Medicine* **79(5)**, 394–406.
33. Philibert, I. (2005) Sleep loss and performance in residents and non-physicians: a meta-analytic examination *Sleep* **28(11)**, 1392–1402.
34. Nelson, D. (2007) Prevention and treatment of sleep deprivation among emergency physicians *Pediatric Emergency Care* **23(7)**, 98–503; quiz 504–505.
35. Spickard, A. Jr., Gabbe, S.G., and Christensen, J.F. (2002) Mid-career burnout in generalist and specialist physicians *Journal of the American Medical Association* **288**, 1447–1450.

Chapter 4
Measuring and Maintaining Faculty Health

Mamta Gautam

Abstract Faculty health in an academic medical center deserves specific and focused attention, as the health of the faculty affects the health of academic peers, students, patients, and the community. The stigma of illness persists in the culture of medicine and serves as a barrier to the seeking of help by faculty members in need. Measures of wellness can be used in the workplace to determine the presence and level of burnout and illness, and encourage the use of resources. The results of such measures can lend support for the creation of a wellness program for the faculty. An effective wellness program addresses aspects of prevention and health promotion, education, intervention, research, and identification of available resources. Specific steps are outlined to facilitate the development of such a faculty wellness program to maintain faculty health.

Keywords Faculty health, stigma in medicine, burnout, faculty wellness program, healthy medical workplace

The workplace is a key determinant of health. While this is true of all workplaces, it is especially true in academic medical centers, where the culture of medicine enables denial of faculty health problems and sustenance of unhealthy behaviors, thereby perpetuating factors leading to an unhealthy medical workplace. The medical faculty train doctors and scientists, and serve as a role model for behaviors within medicine. Further, healthy health care professionals serve as healthy role models for their patients. Encouraging and supporting a healthy medical faculty, therefore, will have a far-reaching positive impact on the health of physicians, their patients, and the community at large.

M.Gautam
Founding Director, Faculty Wellness Program, Assistant Professor, Department of Psychiatry, University of Ottawa, Ottawa, Ontario
e-mail: mgautam@rogers.com

T.R. Cole et al. (eds.) *Faculty Health in Academic Medicine,*
© Humana Press, a part of Springer Science + Business Media, LLC 2009

Stigma in the Culture of Medicine

It is not easy for doctors to admit that they may need help in becoming healthier. The culture of medicine sets high expectations of its clinicians, educators, and scientists. It promotes hardwork, conscientiousness, perfectionism, compulsiveness, and thoroughness. It encourages self-sacrifice and delay of personal gratification. These personality traits and attitudes of physicians were well-described by Gabbard and Menninger in their work with physician couples [1].

Ideal faculty members come in to work early and leave late and are always available. They pay attention to every detail, are careful, highly responsible, reliable, and trustworthy. They are tough, strong, and in control. They can handle anything and everything. They take care of others and are ready to help when needed. This perception is reinforced by their teachers, their training, their colleagues, their students, and their patients. It can lead to misperceptions.

Many in academic medicine believe: "It is wrong to get ill and to need and ask for help."

An Emergency Medicine doctor completed her entire shift although she was having a miscarriage, bleeding heavily, and feeling weak. That day, she had assessed and admitted 3 other women to hospital who were also experiencing a miscarriage, but felt that she needed to continue to work and not irresponsibly burden her colleagues.

The family doctor spent the whole night up with her 9-month old baby, who was rubbing his face into her and crying inconsolably—an unusual state for this baby. He felt hot to the touch and was unable to sleep. She was not sure if the situation was serious enough to take him to the hospital and was relieved when he finally fell asleep around dawn. She felt surprised when she checked in on him in the morning and found him sleeping, with pus and blood draining from his ear. She felt guilty at having made him endure the pain of otitis media.

The senior scientist was still reeling from his visit to the doctor, where he heard he had bowel cancer. He decided to wait and find out the exact pathology and staging before he told his colleagues "so they would believe him when he asked for time off for his treatments."

While the concept of no needs/no help is believed to be true for physical illnesses, it feels even truer for mental illnesses.

A physician suffered an acute myocardial infarction and was hospitalized in the cardiac unit of his hospital. His department sent a card and flowers, and many of his colleagues stopped by his hospital room to visit and offer support and encouragement. At the same time, another colleague from the same department was diagnosed with severe depression and was admitted to the same hospital's psychiatric unit. There were no cards, flowers, or visits for him. The physician with the myocardial infarction remarked, "I had the more noble disease."

A resident spoke to me about his developing depression. The attending physician in charge of his team became angry at him for coming to work late each morning and yelled at him publicly while rounding with the team on the ward. I asked him if the attending physician had asked him if he was well. He replied,

"No." I then asked him if he told the physician about his depression, insomnia, and fatigue, or tried to explain why he was late. "No" he answered, seeming incredulous. "I would much rather he thought I was lazy than know I was depressed."

Another misconception is: "It is weak or lazy not to choose to work all the time or stay home if ill." In line with this idea, many doctors and scientists find it hard to take holidays and often do not take all the time they are allowed to have. It seems to be difficult to go home early the day after call, although the protocol allows so, for fear that leaving early would be seen as signaling a lack of commitment or dedication. Many go to work ill out of a need to show they can still work and "be strong."

Nowhere is the stigma of illness more present than within the field of medicine. A stigma is a stain; a mark of defect, of disgrace, of shame. Colleagues are stained, weaker somehow, if they have a diagnosis of illness. Mental illness is a bigger stain, a bigger disgrace. As a result, doctors may not reach out for help easily, may not attend workshops on stress or health-related topics as they do not want to be seen to need such information. They may not accept prescription medications when advised or will drive to a pharmacy in a distant neighborhood where they feel more inconspicuous to get their prescriptions filled.

Consequences of Stigma

The stigma about illness in doctors can lead to some serious consequences and create barriers to seeking help. It reinforces intellectual defenses and so can prevent us from asking for help when needed. Reaction formation is a common defense for them that leads to giving others all the care and attention they would like to receive themselves. They stay late at the hospital, doing yet another round on the wards, to ensure they have addressed all that was necessary and listened to others' concerns with care and attention. Meanwhile, they deny that they have any problems: *"Everything's fine; I'm all* right." They can minimize problems: *"It's not that bad. After all, I am getting my work done."* They can rationalize problems: *"It's only because I have not had a holiday in a couple of years. As soon as I can get away, I'll be fine."* Such cognitive processes work effectively to make problems seem to go away, thereby alleviating the need to address them. Working harder can serve as a distraction and a way to avoid time to think about the situation. As a result of these unconscious mechanisms, the doctor delays seeking help and so is often much more ill than the average patient when first assessed. Sometimes the doctor chooses self-diagnosis, self-medication, and self-treatment.

Other barriers to seeking help include fear of being exposed or being "found out" as weak or having failed, fears of being judged. There are fears about privacy and confidentiality, fears about the impact on licensing, or on the ability to obtain insurance coverage. There are fears about the lack of control in the treatment process.

In reality, it is not wrong to seek and to receive help. What is wrong is to not get help when it is required. Having a wellness program in place in the academic medical center helps to reduce the stigma, and to normalize the stressors inherent in medicine and encourage strategies to manage them in a healthy manner. While culture change is never easy or quick, it can occur and gains momentum as the benefits become increasingly visible.

Background to the University of Ottawa's Faculty Wellness Program

In 1995, the University of Ottawa set up an innovative and effective wellness program for its faculty after a serious suicide attempt by a highly respected colleague shocked the community. The Dean of Medicine at that time, John Seely, M.D., established a Task Force on Faculty Stress to identify, validate, and define the serious issue of stress among physicians generally, and specifically among those within the Faculty of Medicine at the University of Ottawa. This Task Force held numerous educational events for the faculty to promote increased awareness of the stress in medicine, to identify the scope of the problem, and to address the development of support programs. It presented educational workshops, offered preventive strategies, and helped link faculty members to resources needed. The Task Force led to the creation of the Faculty Wellness Program in 2000 to formalize this program.

A Measure of Our Wellness

Although a specific incident led to the creation of the Task Force, further data were required to define and highlight the need for a formal health service. In spring 1999, the Task Force collaborated with Dr. Linda Duxbury, Professor, School of Business, Carleton University, Ottawa, in her research on workplace stress. She studied five departments within the Faculty—Pediatrics, Psychiatry, Anesthesia, Medicine, and Ophthalmology. An 18-page survey, modified for use by physicians, was sent to all the faculty members of these five departments.

The survey results confirmed a high level of stress. The response rate was 30%. The respondents stated that they worked an average of 59 hours per week and had high job stress with heavy workloads and conflicting demands. There was low job satisfaction in 48% of respondents, and 15% felt there was low job flexibility. Ten percent felt they had low job control. Overall, respondents felt that their work performance was not being valued. Half of the respondents were thinking on a weekly basis about leaving academic medicine. Stress and burnout was found to be higher than in the general population, with burnout present for one in five, anxiety present for one in three, and depression present for one in five of the respondents. Of special note, in the preceding three months, one in four had felt really stressed; one in five had poor emotional health, 12% had thought about suicide, and 7% had planned suicide.

These results were astounding and confirmed the reality of the problem of major stress and difficulty among faculty members. They provided the impetus to move the Task Force from an ad hoc body to a standing and permanent office in the Faculty of Medicine. The move was supported fully by the new Dean, Peter Walker, M.D., and his support led to the creation of the Faculty Wellness Program and its advisory committee.

Other Health Measurement Instruments

While we were fortunate to be able to work with a leading researcher on workplace stress and create a tailored instrument to measure our stress, there are other standardized instruments available. The Maslach Burnout Inventory (MBI), developed by Christine Maslach, is commonly used to assess the level of burnout in the workplace [2]. Burnout derives from chronic overstress, and while not a psychiatric diagnosis, it can lead to many serious diagnoses. In burnout, the demands of the workplace exceed and exhaust the resources available. This situation is becoming a way of life in medicine.

The first dimension of physician burnout is emotional exhaustion, in which the physician manages to function and get through the day, but is emotionally drained at the end. There is little energy left over for anything or anyone else. Those with burnout become more irritable and negative and start to pull away from people who may require anything from them.

The second dimension, depersonalization, finds the physician more isolated from family, friends, colleagues, patients, and increasingly cynical and negative. The third dimension is perceived ineffectiveness such that the physician has lost the earlier sense of satisfaction from work and considers leaving medicine. Burnout can lead to work difficulties and law suits; marital problems; physical problems such as sleep disorders, hypertension, and cardiac disease; alcoholism and drug addictions; and psychiatric illnesses such as depression, anxiety, eating disorders, and suicide.

The Maslach Burnout Inventory has been used by many authors as one of several surveys to assess the health of a group of physicians [3–6]. A 2003 study of physicians in Alberta, Canada, by researcher Robert Boudreau of the Faculty of Management at the University of Lethbridge, measured the prevalence and severity of burnout [3]. Four measures of burnout were used, including the Modified Maslach Burnout Inventory, the Pines and Aronson Burnout Measure [7], the Boudreau Burnout Measure, and the Rafferty et al. Overall Self Assessment of Burnout [8]. These instruments were shown to have very good psychometric properties and are highly intercorrelated. This burnout study showed that 48.6% of the respondents were in the advanced stage of burnout.

In 2002, Gabbe et al. led a study of burnout in chairs of obstetrics and gynecology [6]. A questionnaire was devised and sent to 131 department chairs throughout United States and Puerto Rico. They achieved a 91% response rate and found that 22% of the respondents were very dissatisfied with their positions. This study used the Maslach Burnout Inventory-Human Service Survey (MBI-HSS) and revealed a high subscale score for emotional exhaustion, a moderate-to-high level of depersonalization or cynicism, yet a high score for sense of personal accomplishment. It demonstrated burnout to be more common in new chairs, and in those who had less spousal support.

A similar study was carried out by Saleh et al. in 2007 to examine the prevalence and severity of burnout among academic orthopedic departmental leaders [9]. They developed a questionnaire with a format similar to that utilized by Gabbe et al. and the American Gynecological and Obstetrical Society. Their eight-page prospective survey was divided into seven sections, focusing on demographic information, self-efficacy, identification of departmental stressors, job satisfaction, work–life balance, assessment of features of burnout, and family and spousal support. A final response rate of 69% was achieved. They found moderate-to-high levels of emotional exhaustion, moderate-to-high scores on the depersonalization scale, and a high sense of personal accomplishment. They demonstrated a significant relationship between the self-efficacy score and burnout. As self-efficacy increased, burnout decreased.

The General Health Questionnaire (GHQ-12) is another well-recognized and validated tool for assessing mental health [10]. It uses Likert scoring, or an alternative method, producing scores between 0 and 12 with higher scores indicating a higher probability of disorder. This tool was used by McManus and associates [4] and McLoughlin and colleagues [11] in their studies on physician stress and mental health.

The Stress Arousal Checklist (SACL) is a validated measure to assess stress and arousal levels [12]. It has been used by many authors, including McLoughlin et al. [11]. It is a list of 25 adjectives, which could be either stressors or arousers, and either positive or negative, to describe feelings and moods. Respondents use a four-point scale to indicate how accurately the adjective matches their current state.

In an important study in 2002, Delva and colleagues used earlier research to develop two separate tools to study doctors' workplace and how doctors work [13]. The Approach to Work Questionnaire (AWQ) shows three factors: Surface-Rational, Surface-Disorganized, and Deep, which relate to different methods (independent, problem-based, and consultative) and motivations (internal and external) for continuing medical education. The Workplace Climate Questionnaire (WCQ) shows three dimensions: Choice-Independence, Supportive-Receptive, and Workload. These questionnaires offer a view of the doctor's learning environment and approach to work at a particular point in time. These questionnaires were used as part of the methods by McManus and colleagues [4].

Occupational stress can be measured by a number of other tools. In the Job Stress Inventory, occupational stress is measured by participants rating on a Likert scale the amount of stress caused by 38 possible stressors [14]. Coping strategies can be measured using the COPE, a survey with 15 subscales including active coping, planning, seeking supports, suppression, use of religion, acceptance, denial, disengagement, substance use, and humor [15]. The Occupational Stress Indicator is an 18-item health scale for the assessment of mental health [16]. It makes use of a six-point Likert scale, and a high score is indicative of poor mental health. This scale appears to have good internal consistency. Job satisfaction can be measured by the Warr, Cook, and Wall Job Satisfaction Scale [17]. A study of Irish general practitioners by O'Sullivan used all of the above four scales to measure job stressors and coping strategies to predict mental health and job satisfaction [18].

Finally, as we did in our study at the University of Ottawa, questionnaires to assess health can be developed to address specific needs and aspects of the medical group under study. Another excellent example is described in a study of the health and practice of surgeons by Harms and his associates [19]. In their study, the researchers devised a detailed interview administered directly in person or by phone, to examine demographics, health, surgical practice, and major sources of stress. The researchers enjoyed an amazingly high response rate of 97%, largely due to the personal nature of the contact. In addition to the high level of follow-up, such contact allowed discussion of specific and confidential issues. The only potential weakness of this approach might be an underreporting of symptoms in a personal interview.

Any one or combination of these tools will provide a snapshot of the health of the faculty and deliver the data required to assess the faculty's needs for programs to assist in promoting and maintaining health and well-being.

Setting Up the Program

Faced with statistics suggesting that large numbers of faculty were encountering increasingly high levels of stress, the University of Ottawa launched a program for its medical faculty. In June 2000, the University of Ottawa Faculty Wellness Program was established as an official program of the Faculty of Medicine. It was hoped that this positioning would encourage overstressed physicians and scientists to get help, which they have traditionally been reluctant to seek out. Buy-in from the senior leadership is another essential for a wellness program to succeed. The Dean, the Dean's advisory group, and Chairs of Departments were all supportive of the establishment of the program.

From the outset, the Faculty Wellness Program was recognized as part of the structure of the organization under the auspices of the Associate Dean of Professional Affairs. The Terms of Reference were drawn up and stated that the Faculty Wellness Program was committed to the enhancement of the well-being of the faculty, with comprehensive initiatives in the five areas of Education, Prevention, Research, Resources, and Intervention. The program was approved by the Faculty of Medicine and the Senate of the University of Ottawa. A budget was drafted to support the activities of the Program.

The Faculty Wellness Program Mandate stated that the Program would provide assistance for a variety of difficulties. These included stress and burnout, depression, anxiety disorders, conflict resolution, grief and bereavement, alcohol and drug addictions, relationship issues such as separation and divorce, financial management, time management, career guidance or renewal, and support during litigation or complaints.

Faculty members were actively recruited for a volunteer program committee. The work of the advisory committee was strengthened by broad participation from the departments of the Faculty, taking the program beyond solely psychiatric affiliation. In addition, junior faculty, residents, fellows, basic scientists, and support staff were represented in this group. All committee members were respected leaders and colleagues. Each committee member was provided with a list of contact information for all members of the committee to facilitate networking and support between members. The Committee was chaired by the Director of the Wellness Program.

A Connector Program, established by the Task Force, was revised and maintained by the Program. The Connector Program was similar to an Employee Assistance Program with primary connectors who could assist in gaining access to specialist resources. Five key physicians were identified who would serve as the primary contacts if a faculty member required assistance. One connector at each hospital site was identified. However, colleagues were encouraged to contact any one of the five physicians and were not restricted to the connector at their site.

In addition to the Connectors, a list of specialists was created. The specialists were identified, personally contacted, and requested to agree to be available on an urgent basis. Among the specialists were psychiatrists, psychologists, social workers, lawyers, legal mediators, and financial consultants. The focus of the mental health care workers covered the spectrum from child to adult therapy; individual, family, and marital therapy; and English, French, and other languages of therapy; as well as specialty areas such as bereavement and cancer. Resource personnel were identified for urgent hospitalizations. These were primarily the Chiefs of Psychiatry at local and provincial non-teaching hospitals where there was a greater degree of privacy for academic physicians. The Resource List, which named these specialists, gave full contact information including area of specialty. It was only made available to the five Connectors.

Promotion of the Program

The Program was promoted for access, both internally within the Faculty of Medicine and the University of Ottawa, and externally within the city of Ottawa. Although created at the university center, non-academic physicians working within the entire community were also encouraged to utilize this resource. Also, the Program was promoted at provincial, national, and international levels in presentations at medical meetings.

A logo was created to provide instant recognition for this program. The logo chosen from the artist's suggestions depicts the Caduceus with the face of a physician and one of the two wings folded over to represent self-care. A brochure highlighted the logo, contact information for the Program, the five Connectors and their contact information, and the Mandate of the Program with areas in which assistance was available. The brochure was written in both English and French, and distributed in a mail-out to faculty members.

A web site for the Wellness Program was developed as part of the Faculty of Medicine web site. The web site provided some information similar to that on the brochure. In addition, it offered details on the Connectors and their contact information, and information on the Program, its Committee members, and upcoming committee meetings. A basic stress self-test was offered. The web site contains links to other useful and interesting sites, such as medical organizations, medical leadership groups, humor sites, hospitals, and the medical wellness library collection.

Implementation of the Program

The Program Committee met monthly. In addition, there was regular e-mail contact between committee members and interim meetings, as required. The Director of the Program, as Chair of this Committee, linked back to the Dean regularly to maintain this connection and communication.

Education

Under the mandate of Education, the Program conducted regular presentations on wellness, including warning signs of stress, symptoms of burnout, depression, anxiety, and addictions. Also, information was provided on suicide recognition and prevention. These presentations were primarily delivered during departmental Grand Rounds, retreats, seminars, and workshops. Specific workshops were designed for different departments and conducted on topics such as Stress Management, Assertiveness Training, Balancing Home and Work Lives, Legal and Emotional Aspects of Divorce, Suicide Prevention, and Dealing With the Stress of Litigation. These were well-attended and evaluated.

Prevention

As part of the goal of Prevention, the Neighborhood Watch program was defined, promoted, and maintained. It was based on the concept of encouraging people within a neighborhood to watch out for each other. The initiative used the analogy of witnessing a neighbor having angina while shoveling snow, knowing the signs and symptoms, and the immediate response of giving aid. Similarly, faculty members were encouraged to know the warning signs of stress and distress and to watch out for them in colleagues. The aim was to increase the comfort level in approaching and aiding colleagues with possible concerns.

A Massage Therapy pilot project was suggested, which would offer seated massages within the hospital for physicians to use at a minimal fee. This project prompted discussion and concern about physicians being offered special services. Further action on this project was deferred.

Promotion of regular exercise by faculty members was addressed. A list of existing exercise facilities at the hospital sites was created, circulated, and offered on the web site. The list provided locations, times of operation, and equipments available. A fitness coach was invited to come to hospital sites on specific days to teach and lead Walking Groups. A member of the Running Room staff conducted a running clinic and started a group of colleagues running on a regular basis.

Promotion of caring for physical health was also begun. Faculty were regularly and openly encouraged to obtain their own family physician. A list of family physicians who were comfortable treating colleagues was made available.

Mentoring Programs were encouraged and supported. Mentors were matched and assigned to each member of the faculty and students. The mentor groups were encouraged to be available to deal with both academic and psychological issues, as required. Presentations on mentoring were organized to assist mentors to learn new skills, to support their peers, and to share effective mentoring strategies. A Buddy System was put in place for all medical students to provide initial support. Mentorship Guidelines for the Medical Student Mentor Program were developed to strengthen the mentor program and ensure that the mentors would be better available and able to offer resources.

Planning was initiated on a full-day conference to focus on the Humanities in medicine. Focus will include the arts, theater, music, humor, and ethics. The goal of this conference is to provide a forum for colleagues to learn, share, and express how these elements enhance their ability to take better care of themselves, as well as their patients.

In addition, the Program served as a point of contact to enable colleagues to come together to start social groups. Examples include a Book Club, a Movie Club, and a Music Appreciation Group. Once these groups were formed, their members were left to maintain the group without further involvement of the Wellness Program.

With the thought that recognition and positive reinforcement are key ways to improve and maintain morale, programs were devised to ensure that the Faculty of Medicine promoted regular recognition of its members. A complete list of department members was created and updated to support a program to acknowledge milestone years at the university, with specific tokens of appreciation. An "Unsung Heros" wall was created, which was a bulletin board in a conspicuous location in each site that highlighted achievements and promoted major and minor personal triumphs. Colleagues were encouraged to nominate peers for awards. The Dean's newsletter regularly contains praise and recognition of faculty members. Each department was encouraged to sign birthday cards for its members, thereby giving them an annual reminder of how much they are appreciated by peers. The Wellness Program led the initiative to have the mayor of the City of Ottawa proclaim an annual day as Physician Appreciation Day (Inaugural Day, October 16, 2001), to publicly acknowledge and appreciate the multiple roles of physicians in the community.

Departmental retreats were created and conducted. The Wellness Program was instrumental in helping departments organize retreats during which they set up their own departmental wellness programs. They were assisted in identifying problems causing stress in their department, learning how to address such issues, how to create a departmental wellness mission, and encourage mutual support and care. Each department who held such a retreat was encouraged to identify a departmental ombudsman who is interested in wellness issues, is trusted by colleagues, and is approachable. This ombudsman was given further training in wellness and was available to assist with problems as needed.

Research

Focus groups were held within several departments to explore sources of stress and possible solutions. The Chair of a given department was initially present to bring the group together and voice support, and then left the room to allow open discussion. The group was asked four questions:

1. What works in your department?
2. What does not work in your department?
3. What is the level of morale in your department?
4. Any specific suggestions that could make things better?

Information from the focus groups was collated and presented to the departmental chairs in written and oral manner. This process led to the recognition that the departmental chairs shared common challenges. One of the responses was the creation of two retreats for the Dean's Group and departmental chairs (2001 and 2002). The focus of these two annual retreats was leadership challenges and wellness. The Faculty Wellness Committee recognizes that there is much more to be learned about its subject and plans to identify further areas of research.

Resources

The Faculty Wellness Library was created as a virtual collection within the Medicine library. It provides a list of resources on wellness issues, including articles, books, videos, and web sites, and is regularly updated by the librarian. The local fitness facilities are listed and offered for faculty. The Program maintains and updates the list of the Connectors and the specialist resources. The Program also serves as a resource for departments seeking an educational seminar or workshop tailored to their needs.

Intervention

The Program received an average of five calls for advice or assistance per week. In Year 1, there were 263 calls, and in Year 2, there were 212 calls received and handled. This number was much more than anticipated. Of the total of 475 calls in the first 2 years, 43% were from male faculty; 57% from female faculty members. With regards to medical specialty, calls were from Family Medicine (23%), Medical Specialists (76%), and medical scientists (1%). Eighty-six of the calls were for problems related to self, 9% for problems dealing with a child, and 5% for problems dealing with a colleague.

The nature of the problem, as assessed by the initial contact person, was:
Depression and anxiety 36%
Stress and burnout 28%
Substance abuse 9%
Relationship problems 8%
Litigations/complaints 8%
Conflict resolution-work 7%
Career guidance/renewal 3%
Time management 1%

There were 12 additional calls—6 from residents, 4 from medical students, and 2 from members of the support staff. These were handled similarly to the other calls and made us aware of the need to expand the program to serve these groups.

Steps to Create a Wellness Program

1. Document the need, through a survey or measurement instrument.
2. Ensure essential buy-in from the top. The Dean, Department Chairs, Hospital Chief of Staff and CEO must openly and visibly support a Wellness Program.
3. Ensure that it is embedded in the academic infrastructure. Obtain formal university faculty and senate approval, and a place on the organizational chart.
4. Define a budget and identify needed financial, administrative, and personnel resources to be dedicated to the Wellness Program. Finances are required to offer the resources and to compensate for time spent by the Director, Connectors, and others who give their time and expertise to the Program. Also, committed administrative assistance is required to maintain this Program and enable regular availability to answer to colleagues who request assistance and connection to resources. Funding for specific training in dealing with crises is required, particularly if the Program is to be led by a non-mental health care professional.
5. Create terms of reference and a mandate for the program.
6. Ensure a full range of participation from all aspects of the faculty on the program committee to represent all needs—all specialties, junior faculty members, medical students, residents, minority groups, clinicians, researchers, and scientists.

7. Prospective committee members are ideally invited by the Dean to join, thereby enhancing the support and validity of this Wellness Program.

8. Recruit respected leaders to champion the cause and de-stigmatize illness.

9. Recognize that a Wellness Program is needed and will likely be well-utilized if implemented. "If you build it, they will come". Ensure a full complement of resources to response to identified needs.

10. Ensure that the process to disseminate information from the Wellness Program, to each member of the faculty is in place and is effective. A comprehensive faculty e-mail list needs to be created as well as a system for ensuring that Chairs and their assistants pass along information to all department members regularly and consistently.

11. The location of the workshops should be carefully chosen to preserve privacy. For example, the workshop on Divorce was poorly attended as it was located in a highly visible room in the Dean's offices' area. Some likely attendees later told us that they did not attend because of lack of privacy.

12. The Director of the Program requires support and resources and cannot be expected to administer the program, provide rounds and educational sessions, and also provide clinical care to colleagues. It is essential that the administrative and the clinical work be done by separate people.

The current health care system is ailing and has a negative impact on the wellness of physicians. Any highly functioning healthy individual, placed in an unhealthy environment, can become unhealthy. We cannot entirely change the health care system. However, we can establish a supportive work environment within this system to allow for healthier medical faculties and physicians.

References

1. Gabbard, G.O., and Menninger, R.W. (1988) *Medical Marriages* Washington: American Psychiatric Press, 23–34.
2. Maslach, C., and Leither, M.P. (1997) *The Truth About Burnout* San Francisco, CA: Josey-Bass
3. Boudreau, R.A., Grieco, R.L., Cahoon, S.L., Robertson, R.C., et al. (2006) The pandemic from within: two surveys of physician burnout in Canada *Canadian Journal of Community Mental Health* **25(2)**, 71–88.
4. McManus, I.D., Keeling, A., and Paice, E. (2004) Stress, burnout, and doctors' attitudes to work are determined by personality and learning style: a twelve year longitudinal study of UK medical graduates *BMC Medicine* **2**, 29.
5. Bruce, S.M., Conaglen, H.M., and Conaglen, J.V. (2005) Burnout in physicians: a case for peer-support *Internal Medicine Journal* **35**, 272–278.
6. Gabbe, S.G., Melville, J., Mandel, L., and Walker, E. (2002). Burnout in chairs of obstetrics and gynecology: diagnosis, treatment, and prevention *American Journal of Obstetrics and Gynecology* **186**, 601–612.
7. Pines, A., and Aronson, E. (1988) *Career Burnout: Causes and Cures* New York: Free Press.
8. Rafferty, J.P., Lemkau, J.P., Purdy, R.R., and Rudisill, J.R. (1986) Validity of the Maslach burnout inventory for family practice physicians *Journal of Clinical Psychology* **42(3)**, 488–492.

9. Saleh, K.J., Quick, J.C., Conaway, M., and Sime, W.E., et al. (2007) The prevalence and severity of burnout among academic orthopedic departmental leaders *Journal of Bone and Joint Surgery. American Volume* **89**, 896–903.

10. Goldberg, D. (1992) *General Health Questionnaire (GHQ-12)* Windsor: NEFR-Nelson Publishing.

11. McLoughlin, M., Armstrong, P., Byrne, M., and Heaney, D., et al. (2005) A comparative study on attitudes, mental health, and job stress amongst GP's participating, or not, in a rural out-of-hours co-operative *Family Practice* **22**, 275–279.

12. Corcoran, K., and Fischer, J. (1987) *Measures for Clinical Practice: A Sourcebook* London: Free Press.

13. Delva, M.D., Kirby, J.R., Knapper, C.K., and Birtwhistle, R.V. (2002) Postal survey of approaches to learning among Ontario physicians: implications for continuing medical education *British Medical Journal* **325**, 1218.

14. Cooper, C.L., Rout, U., and Faragher, B. (1989) Mental health, job satisfaction, and job stress among general practitioners *British Medical Journal* **298**, 34–35.

15. Carver, C.S., Scheier, M.F., and Weintraub, J.K. (1989) Assessing coping strategies: a theoretically based approach *Journal of Personality and Social Psychology* **56(2)**, 267–283.

16. Cooper, C.L., Sloan, S.J, and Williams, S. (1988) *Occupational Stress Indicator Management Guide* Windsor: NFER-Nelson.

17. Warr, P., Cook, J., and Wall, T. (1979) Scales for the measurement of some work attitudes and aspects of psychological well-being *Journal of Occupational Psychology* **52**, 129–148.

18. O'Sullivan, B., Keane, A.M., and Murphy, A.W. (2005) Job stressors and coping strategies as predictors of mental health and job satisfaction among Irish general practitioners *Irish Medical Journal* **98(7)**, 199–202.

19. Harms, B.A., Heise, C.P., Gould, J.C., and Starling, J.R. (2005) A 25-year single institution analysis of health, practice, and fate of general surgeons *Annals of Surgery* **242(4)**, 520–526.

Part III
Personal and Social Dimensions

Chapter 5
The Architecture of Alignment: Leadership and the Psychological Health of Faculty

Susan H. McDaniel, Stephen P. Bogdewic, Richard L. Holloway, and Jeri Hepworth

Abstract Faculty health and productivity depend on the alignment between an academic medical center's mission and goals and that of its faculty. Funding difficulties, multiple roles, and diverse faculty contribute to what can be a very stressful environment. The psychological experience of faculty in academic medical centers depends on the individual faculty's temperament, skills, and past experience. It is also mediated by the institutional culture, which is heavily influenced by leaders' ability to articulate vision, utilize emotional intelligence (EI), and develop trust. This chapter examines the links between faculty and their institutional culture, and emphasizes the role of leadership in facilitating alignment to improve faculty and institutional health. Tables are provided to assess both institutional and individual health to enable faculty to work towards alignment of their goals with that of their chosen institution.

Keywords Leadership, psychological health, alignment, administration, faculty support

Academic medical centers are built on the fundamental interplay between the faculty who make them up and the structures that support their activities. The success

S.H. McDaniel
Professor and Associate Chair, Department of Family Medicine,
Director, Institute for the Family, Department of Psychiatry, University of Rochester Medical Center, Rochester, New York, USA
e-mail: Susanh2_McDaniel@urmc.Rochester.edu

S.P. Bogdewic
Associate Chair and Professor of Family Medicine, Executive Associate Dean for Faculty Affairs and Professional Development, Indiana University School of Medicine, Indianapolis, Indiana, USA

R.L. Holloway
Associate Chair and Professor of Family and Community Medicine, Associate Dean for Student Affairs, Medical College of Wisconsin, Milwaukee, Wisconsin, USA

J. Hepworth
Department of Family Medicine, Office of Faculty Leadership, University of Connecticut School of Medicine, Farmington, Connecticut, USA

T.R. Cole et al. (eds.) *Faculty Health in Academic Medicine,*
© Humana Press, a part of Springer Science + Business Media, LLC 2009

of the architecture of academic medical centers is highly dependent on the psychological health of the faculty who populate them. How the health of the academic medical centers and its faculty interact is the primary subject of this chapter, beginning with two illustrative cases:

Jane T is a family physician who just finished her postdoctoral fellowship, including a Masters in Public Health. She plans a career as a clinical researcher, developing interventions to improve asthma outcomes for African refugees whom she serves as a primary care clinician. Her salary is 50% clinical (with an expectation of 2000 outpatient visits/year and 4 weeks on-call in the hospital), 10% teaching (with an expectation of teaching 4 seminars/year and precepting residents once/wk), and 30% protected time to plan and collect pilot data for a federally funded research Career Development K-award. Jane and her husband would like to start a family. She enters academic medicine with equal amounts of idealism and ambivalence—wondering if she will be able to juggle clinical demands and research, teaching and learning, work and family life.

Fred S is a family physician who serves as Jane's mentor. He is a tenured Full Professor. He has many publications about assessing and intervening with mental health problems in primary care behaviors, including several in JAMA *and the* New England Journal. *His research mentees, like Jane, find him an excellent mentor; however, he has difficulty getting along with his colleagues who view him as non-collaborative and entitled. Fred's moods follow his funding, which is intermittent. His current funding is 40% research grants, 30% clinical (20% out-patient 10% attending on the inpatient unit), and 30% education (including mentoring, teaching faculty development seminars for junior faculty, and teaching residents). While Fred is a talented and well-recognized clinical researcher, he finds what he calls "chasing funding" very stressful. He is perpetually writing grants, to stay ahead of funding cycles. When research grants dry up and that portion of his salary must be replaced with clinical work, patient care leaves little time to write grants or papers.*

The psychological experience of faculty in academic medical centers in the early 21st century is filled with stimulation and stress, responsibility and reward, and change and contradiction. The experience is then embedded within an American healthcare system that is chaotic. Further, it is crumbling under the weight of sky-rocketing costs and the lack of a central organizing structure.

Faculty in academic medical centers are generally dedicated to their work as clinicians, educators, and scholars, even as they are aware of discontent expressed by physicians in practice [1]. The challenge is to manage their diverse roles during difficult times and align their own individual goals with the collective mission of the academic medical center. That alignment may enhance the unique joys they hoped to experience—touching clinical moments, joys in research and teaching, and the chance to work with brilliant colleagues. To the extent faculty are aligned with the larger system's mission, they have the opportunity to succeed. To the extent they are misaligned, faculty will experience greater levels of stress.

Each faculty member responds to these challenges with a unique coping style born of personal history, temperament, and learning. The psychological experience of faculty in academic medical centers is as diverse as their ages, genders, disciplines, and ethnicities. Few research studies exist about the psychological experience of faculty in academic medical centers, so little can be reported from an empirical base. However, from our collective decades of experience as administrators and faculty in such centers, we have identified common themes to provoke conversation and future research in this important area of faculty health.

This chapter will describe the successes and the stresses common to the experience of diverse faculty in equally diverse academic medical centers. We focus on factors contributing to the psychological health of the individual faculty member and the health of the institution, and to leadership as the pivotal link in aligning the faculty and the institution. We report the few studies that provide data about these factors. We conclude with suggestions for a self-assessment of the individual faculty's psychological health in the workplace *along with* a self-assessment of the psychological health of the institution. We begin with a description of the current landscape and our theoretical framework.

Understanding the Current Landscape

Funding difficulties in today's healthcare environment heavily influence the psychological experience of faculty (As discussed later, the response of leadership to that stress is at least as important.) Federal attempts to contain rising health care costs have resulted in institutional demands for increased clinical productivity without changing the need for faculty to excel as medical educators and scholars. Academic medical centers in the United States reflect the paradoxes inherent in the larger US health care system: cutting-edge high-tech interventions alongside embarrassingly high rates of infant mortality and teen pregnancy, a desire by those with health insurance for easy access to specialists alongside a large number of Americans who have no health insurance and no access, training that emphasizes the physician's legal responsibility for patient outcomes alongside healthcare that now emphasizes team approaches to outpatient and inpatient care, and a system that rewards federally funded research above all else alongside requirements for faculty to care for the sickest of patients and train the physicians of the future.

While it would be tempting to think of the paradoxes present in the academic medical center as applying solely to clinicians, there are parallel challenges for those whose careers are dedicated primarily to discovery. At its most basic, increasing numbers of researchers have their homes in clinical departments, a position providing a daily reminder of the applications of their scholarship. A colleague of one of the authors now scrubs in with his anesthesiology colleagues in the O.R., even though his research comprises basic studies of consciousness in rats. Moreover, the establishment of multidisciplinary centers has extended the trend toward placing the basic researcher in an increasingly applied context. And finally, the recent clinical translational research initiative of the National Institutes of Health (NIH) demands that centers assemble clinical and research faculty as collaborators with a goal of application to practice rather than pure discovery. Hence, the paradoxes faced by clinicians are faced in near mirror image by researchers: While institutions declare alignment with a variety of scholarly priorities, it may well be that the "true" institutional mission has more to do with becoming more and more eligible for a wider base of funding, including that procured by scientists translating their research to clinical applications.

Then there is the increasing diversity of the American population. Medical schools on the whole are making progress, but not keeping up with changing demographics. For example, minority populations in the United States were estimated at 27%, whereas the graduating class of 2006 had 6.9% minority students [2, 3]. Medical school faculty remain predominantly white and the leadership predominantly male. Similar to the students, only 7.2% of medical school faculty are minority, according to the AAMC [4]. While the women often surpass men in number in medical school classes (50.4% of students nationally graduating in 2006 were women), they are underrepresented on the faculty of our academic medical centers (32.2% in 2006) and rare in leadership roles. In sum, academic medical centers do not look like the neighborhoods within which they reside. Tension around issues of diversity, either dealing with differences or dealing with the pressure to diversify, is ongoing for faculty in the early 21st century.

Diversity among the disciplines of faculty in the academic medical center can be another source of stress in the system [5]. For example, Ph.D. researchers compose 26% of the faculty at many academic medical centers, over half of whom work in clinical departments. Most of these faculty are researchers and educators without the possibility of clinical funding. They live and die by grants and a small amount of educational money. The experience of the non-M.D. in a medical center is variable and depends upon the respect (or lack thereof) they earn and receive from leadership.

Our understanding of how the individual faculty member deals with these many challenges is rooted in ecosystemic theory [6]. Ecosystemic theory focuses on individual and group functioning in the context of larger systems. In academic medical centers, systems include the division, the department, the medical school, the medical center, and the larger US healthcare system. Ecosystemic theory draws from Bowen family systems theory, Engel's biopsychosocial theory, and Brofenbrenner's social ecology theory, and assumes a reciprocal interaction between the individual or small group and a larger system [7–10]. This reciprocal interaction is a theme we will develop further in this chapter. In addition, with regard to the importance of leadership in affecting the psychological experience of faculty in academic medical centers, we draw on Friedman's application of Bowen theory to leadership in the workplace [11]. Ecosystemic theory provides a framework for understanding the alignment of the individual faculty with the academic medical center, and of the academic medical center with the larger social context.

The Individual Faculty in the Larger Academic Health Center

The Funding Base of the Academic Medical Center

Data collected since the 1960s find that research funding has remained relatively stable at about one third of revenues generated at academic medical centers, but growth in clinical income has been enormous. Thus, more and more hiring decisions may be linked to a center's higher dependence on clinical revenues. This change has not been gradual; it has been precipitous. Only 6% of operating revenues of medical schools were due to clinical sources in the 1960s; that number leapt to nearly 50% in the 1990s [4]. There appears to be a leveling-off trend today such that the presence of direct clinical revenue, unheard of in medical schools a short time ago, has undeniably become a lifeblood source of revenue for the modern academic medical center.

Hiring expectations and academic advancement may be at odds with one another. Some clinical departments, for example, structure faculty contracts solely around clinical expectations with an understanding that faculty will figure out a way to be academically productive within the context of their clinical responsibilities. Such an embedded understanding of institutional priorities may not be clear until faculty begin to consider their viability regarding promotion.

Cohort Effects

One cannot overlook the impact of generational change on the nature of the faculty entering our academic medical centers today. New faculty will soon be recruited from "Generation Y," those 72 million individuals born between 1977 and 1994. Expectations, attitudes, work ethic, and world view are dramatically different in this generation of emerging faculty from those held by those of us who are hiring them [12]. And, their views of the academic medical center are profoundly different from traditionally held expectations of faculty for academic productivity and support. Perhaps the shift in perspective only reflects the passing of time, as new faculty members have nothing with which to compare their current expectations. They have only known the highly clinically competitive academic medical centers of today. As new faculty meet with old, however, the differences become more apparent.

A recent academic fellowship seminar may illustrate potential emerging generational themes. The very subject matter of the fellowship is "professional academic skills" and preparation for success as a faculty member. All of the fellow attendees have declared their intentions to pursue academic careers and are pursuing a Masters Degree over 2 years. With fifteen years of teaching the seminar, we have seen changes over time in the approach fellows take to their emerging academic careers. However, the teaching team always presumed a fundamental commitment to traditional understandings of academic work. A recent discussion among the fellows provides a dramatic example of how expectations have changed from Boomer to Generation X to Generation Y [13].

When we began the presentation about academic expectations, the room fell eerily silent. "How many of you plan to pursue an academic career," we asked of the seven or so fellows in the room. Not one hand was raised, a first in our experience. "To be honest," replied one assertive internist in the group, "I can't see the benefit in it. I'm very involved in my professional dancing career now, and I have to get the kind of position that will allow me summers off to perform. I thought an academic career was going to give me that kind of flexibility, but what you're telling me sounds just way too demanding!" We appealed non-verbally to the rest of the group. Heads nodded vigorously in assent with the internal medicine fellow. It was an unprecedented turn in a previously relatively predictable educational experience.

The financial changes noted previously, when coupled with the seminar example, provide some beginning insight into the changes in expectations of younger faculty and of the generational changes within the institutions themselves. Historically, medical schools served as sites for indigent care, using such populations as "educational material" for students and residents, as well as a source of research populations. Today, many medical schools have multimillion dollar practice plans and function as fully competitive members of the health care community (One academic medical center in the upper Midwest, for example, is its state's largest practice group.)

Promotion and Tenure

Promotion and tenure systems throughout academic medicine struggle to align their decisions with the evolving nature of the mission of the academic medical center. Most have chosen to create alternative promotion and tenure pathways that will recognize and reward clinically productive faculty as well as those who follow more traditional academic expectations. However, they have generally failed to address the fundamental mission of the school, opting instead to hope that faculty will be able to handle an increasingly intense burden of clinical and scholarly responsibilities, while adhering to criteria that may be out of phase with the goals of the school itself. Little wonder that a harried faculty member, a trauma surgeon, announced he was "trying to get my portfolio ready for rank and tenure review, but most of what I do—the time I spend in the O.R., for example— won't even be considered! Don't you think the school is giving us, you know, kind of a mixed message?" Yes, we do. And such disparities can result in increased stress, anxiety, and uncertainty, a prescription for psychological difficulties.

Some faculty members enter academic careers unaware of the many demands that may be placed on them. Some academic medical centers have Offices of Faculty Development or seminars teaching various academic skills to address the faculty's often skewed perceptions of the demands that will be placed on them institutionally and personally. In the seminar described above, we used an icebreaker that asked faculty to describe their motivations for an academic career. To a person, each faculty member described a fantasy of imparting wisdom to eager learners, thinking deep thoughts, perhaps relaxing in a bucolic atmosphere. Such environments rarely exist in the competitive playing field that is the academic medical center.

Academic medical centers tend to promise less and less when it comes to tenure. Some have rejected tenure altogether while others guarantee very little financially for their tenured faculty. Even so, when the review for promotion and tenure occurs, it tends to adhere to traditional rigor with only grudging regard for the massive changes in the system. When a faculty member is turned down for advancement in rank, the results can be an intensely personal experience of rejection. A *curriculum vita* is, after all, one's "academic life." Try as we may to view the record as something outside ourselves, reviews are personal, with personal consequences.

After a recent promotion and tenure cycle, several candidates were asked to present a panel discussion of the experience to their colleagues in a primary care clinical department. Two of the three were promoted on the first try; the third had been turned down once and then promoted to associate professor two years later. Her words rang poignantly through the room: "I can't tell you what a negative experience it was for me," she began. "I felt as if the whole school had turned against me after the effort I had given. I short-changed my family, my personal life, myself and then ... this. It took me the full three years to decide whether it was worth it to try again. I'm not sure it was, even though I was 'successful' this time around."

While such a dramatic declaration may be unusual, the turn of events reflected in the faculty member's comments encapsulate some of our worst, often unfounded, fears of the promotion and tenure process. The promotion process may often seem completely beyond the control of participant faculty. For those who believe that it is, for the most part, a rational and data-driven process, the limits of anxiety are usually within the normal range. However, for those for whom the process becomes a specter of their loss of control, perhaps anchored in other personal tensions, psychological distress may result.

Psychological Distress

When individuals work in a system that is unpredictable and uncertain, personal distress may well result. High levels of depression, anxiety, and job dissatisfaction— especially in younger faculty—raise concerns about the well-being of academic faculty and its impact on trainees and patient care. Increased awareness of these stressors should guide faculty support and development programs to ensure productive, stable faculty [14].

The academic medical center is often not a particularly hospitable place for those in psychological distress. Note the use of the word "hospitable?" Ironic, isn't it, that the home of hospital affiliates and health care may be a place where it is not particularly wise to have psychological difficulty. In a recent study of medical students' help-seeking behavior, Holloway and Butler discovered several perceived barriers to students' receiving mental health services [15]. While the study focused on medical students, the results may be illustrative of the academic medical center's environment regarding mental health. Students who experienced psychological distress reported that they were less likely to receive help than their symptoms would indicate they need. Forty percent of medical students reported some symptoms of psychological distress, but only 25% of those who were depressed received any help. Fifty eight percent sought help but did not continue, and 16% never considered seeking help. For those reporting anxiety symptoms, the results are even worse: only 17% sought help successfully, 48% sought it but did not continue, and fully 33% never considered seeking help. Clearly, even when students may be distressed, they shy away from seeking help. A study by Givens and Tjia found very similar results [16].

While there may be many personal factors that account for these results, the environment of the academic medical center is certainly among them. The impediments to seeking help include a reported lack of time, a firm belief in self-reliance, concern over documentation of sessions, cost, confidentiality, and the potential impact on one's career. In particular, anxious students doubted the effectiveness of therapy. Depressed students were most concerned about the documentation of sessions. For students who reported both anxious and depressed symptoms, fear of hospitalization against their will was an additional concern.

These are abundant reasons for medical students to avoid seeking help; such reasons are only likely to increase among faculty, particularly concerns about stigmatization and the significance of self-reliance. In a recent, informal survey of research mentors, over half stated that they would not comment on any evident psychological or interpersonal problems with their faculty researcher mentees as they felt that doing so would be "too intrusive." The mythology might well be that faculty know how to take care of themselves; reality speaks to the contrary. Those who are trained to be caregivers are often the least likely to know how to care for themselves, or more pertinently, surrender the necessary control to someone who might care for them [17].

Resultant "problem" behaviors may stem from unmet psychological needs. It is possible, for example, that the productive faculty member who experiences overwhelming psychological distress may manifest such behaviors by being disruptive, bullying, or even abusive. Academic medical centers are a rich canvas on which to paint dysfunctional behavior.

A recent event in a gynecological operating room serves as illustration. Students and residents scrubbed in for a complex procedure that lasted most of the day. While the operation moved along successfully, there was ample time for conversation, banter, and worse. During a downtime in the procedure, a student began describing a recipe for chicken she had attempted the night before, a recipe that involved pounding the chicken to tenderize it. A particularly frazzled surgeon began pressing her on "beating the meat" in reference to the innocuous comment she had made moments before and an obvious additional reference to sexual self-gratification. So extreme and persistent was the surgeon's behavior that the student felt harassed. Further investigation revealed that the surgeon had a longstanding untreated bipolar pattern that was addressed only upon threat of dismissal.

Faculty members who work in intense, isolated surroundings with limited social contact are particularly at risk for such aberrances, though not all result in external displays such as the example above. In fact, one might argue that externalization provides a more readily identifiable expression of distress. Of at least as much concern is the faculty member who internalizes problems and experiences symptoms that may go unobserved.

A senior faculty member in psychiatry, having worked for over twenty five years with the same chairperson, declared himself a candidate for the chair once his leader had retired. The faculty member had been productive, writing books and articles in a narrow but important field of study within psychiatry. However, he was not an appropriate candidate for a major leadership position, especially given his allegiance to a small faction within the highly fractious department. He announced his candidacy as an attempt to garner the support of a disenfranchised group within the department. The search committee became aware of the politicized nature of his candidacy and had no interest in further disrupting the search process by endorsing an unqualified candidate simply because he was an insider. Because he was introverted by nature, no one noticed that the faculty member began hoarding drug samples and food in his office for days, several weeks after his candidacy for the chairmanship had been rebuffed.

When a faculty member who is predisposed to problems finds that she does not fit within the academic world, serious complications may result. Only careful attention and a compassionately assertive intervention saved the faculty member above from serious self-harm.

The person who was eventually chosen as chair of the department, sensing the need for an especially direct confrontation, held the past reality of the failed candidacy and the future opportunity to be a productive member of the faculty clearly in front of the troubled professor. In addition, the Chair made the vision and goals of the newly constituted department quite clear, and offered this person a role in the restructuring. These two faculty remain productive colleagues to this day, some ten years after the event.

While our tendency may be to focus on the individual, the institution and its policies and behaviors may bear closer scrutiny as well.

Institutional Health in the 21st Century

An academic medical center is comprised of multiple missions, multiple constituents, multiple funding sources, and multiple priorities, all of which converge to form an unprecedented degree of complexity. The adage: "Once you've seen one academic medical center, you've seen one academic medical center" aptly describes the variability of organizational and administrative structures of academic medical centers. In turn this variability mirrors the variability of the backgrounds, profiles, and expertise of those who populate the systems. Coupled with the intrinsic uniqueness of each system are the ongoing changes within all health organizations as they clamor to respond to the conflicting pressures for economy, efficiency, safety, patient accountability, and population care.

How is the average faculty member prepared for managing this complexity and never-ending change? Faculty have pursued training in clinical and/or research skills. If fortunate they may have also received some basic preparation for their teaching roles. Beyond that there seems to be an unspoken belief that by simply injecting bright scientists and clinicians into the academic medical center, the faculty will have the necessary capability to master the organizational complexity. Such a mythical belief is a high stakes gamble.

One way that some faculty learn to deal with this complexity is to find a niche and remain bounded by it. This strategy can provide some stability in the life of an individual, and indeed the traditional practices of academic communities support and reward such behavior. But the nature of academic work is changing. Collaboration and its resulting synergism are becoming more the norm. Over the past decade the major focus on quality and safety has demonstrated that health care is not delivered by an individual, it is delivered by a system [18]. Strategies such as team-based learning are becoming more common in medical education. The NIH Roadmap Initiative for research and the development of CTSAs also call for fostering collaborative relationships. It has become less and less possible for faculty to insulate themselves from the complexity of their environment. Without mechanisms to help faculty manage such complexity, the major changes in the culture of academic medical centers can take a considerable toll on faculty well-being.

Faculty development programs and activities focus on strategies for faculty to manage the complexity of academic medicine. Programs cover broad areas such as career development, teaching skills, leadership development, and foundational research skills. There is, however, also great variability in the existence of such programs within academic medical centers. It was not until 2007 that the AAMC established the Group on Faculty Affairs (GFA), a group whose mission is to build and sustain faculty vitality in medical schools and teaching hospitals. While the efforts of faculty development are critically important to managing talent, they alone cannot create the alignment needed to ensure faculty success.

Leadership: Architects of Alignment

What mechanisms exist for creating a healthy environment? Whose responsibility is it to assure faculty that such mechanisms are in place in complex academic organizations? The very nature of being a professional compels an individual to accept a significant degree of responsibility for accomplishing the aims of the profession. But for those in leadership positions, a different degree of responsibility exists. Leaders, by virtue of the values they adhere to and the practices they establish, form, and shape the culture of the organization. Organizational culture then influences all aspects of organizational life. That culture can range from a blame culture (avoid conflict; practice self-preservation at others' expense), to a live and let live culture (do not fix what is not broken), to a culture where each individual can make a meaningful contribution and be valued for it.

This last perspective—the culture of individual contribution within the larger vibrant whole—is the articulated goal of many academic medical centers. As mentioned earlier, the development of this kind of psychologically healthy culture requires attention to the alignment between individual and institutional priorities. If done well, the development of this culture can mediate the exacerbation of the relationship between individual faculty stress and larger systemic stress. When individual and larger system goals are in alignment and recognized, role strain and psychological distress are decreased. To engineer such a culture in a complex organization requires significant ability for leadership.

Leaders of academic medical centers traditionally have been selected from faculty who are recognized as excellent scientists and clinicians, sometimes with good organizational or time management skills.

> Administrators of medical schools and large complex departments are often chosen on the basis of their accomplishments in research. They are generally not as well prepared as their counterparts in industry with responsibility for controlling multimillion dollar enterprises. Furthermore, the organization and management of educational institutions are exceedingly complex; they differ in many ways from the operations in the business sector [19].

If leaders of academic medical centers were ill-equipped for dealing with the complexity in which they found themselves over two decades ago, are present-day leaders any more prepared to manage the level of chaos that exists today? No wonder that academic medical centers throughout the county are seriously struggling to operate with greater efficiency and improved productivity in an environment of declining resources and morale.

With many approaches to leadership available, we find it useful to adapt the core premises of Bennis and Goleman as a guide for leaders who are committed to simultaneously enhancing the organizational health of an academic medical center and the psychological health of the faculty and staff [20, 21]. The core premises for health-promoting leadership include the ability to establish and uphold a vision, to make use of emotional intelligence, and the capacity to engender trust.

Vision

The key to establishing congruency and consistent messages within an organization lies in the ability to align expectations and resources with both individual and organizational goals and objectives. Without such alignment faculty will be pulled in multiple directions, will receive contradictory messages, and can easily become discouraged and possibly burnt out from trying to compensate on the individual level for a misaligned system.

For the leader, alignment between the individual and the system begins with self-knowledge. For example, a leader must believe that an ideal such as professionalism is not just a nice attribute to have within the organization. She must believe that the manner in which colleagues interact with patients, learners, and each other has a discernable impact on the entire mission of the organization—the quality of care, the mastery of teaching, the creativity and productivity of scholarship, and the well-being of all members of the unit. She must then have a keen understanding of how to align an organization so as to create a culture that has a clear purpose, one that clarifies expectations, recognizes contributions, and stays fixed on core values. Her own core beliefs are where the vision starts.

Increasing alignment exists with regard to revenue generation within the academic medical centers. Numerous elements are aligned to ensure that revenue streams are protected. These elements include relative value unit (RVU) systems, compensation policies, revenue sharing plans, partnerships, and so on. However, aligning only one portion of the mission leads to the tail wagging the dog. When clinical productivity is the only mission that is directly rewarded, for example, systems will find limited enthusiasm among faculty for the research or education missions. Even when those three traditional missions are all recognized, the unrewarded goals of colleagueship, collaboration, or citizenship may not be accomplished. Specific leadership strategies must support congruence between an academic medical center, an organizational division or department, and individual faculty.

Resources can be made available to help divisions create their mission statements and goals such that they are aligned with the goals of the larger system. After the mission statements are created, they must then be shared among other leaders and those that work within the unit. Individual faculty should be mentored to also create their own individual development plans, including vision or mission statements. Special attention should be given to identifying strategies that enhance achievement of individual and unit goals. Formal periodic evaluation of faculty should address consistency between stated and achieved goals, as well as congruence between individual and unit missions. Ultimately, the challenge of aligning people and missions in complex organizations is one of communication. People at every level of the academic medical center must regularly be engaged in meaningful discussions of competing missions and leaders must help them discover how to align these missions with individual and institutional core values and beliefs.

Mentoring division leaders should include mechanisms for communication of unit goals with other unit leaders, and periodic larger group reflection on achievement and congruence of unit goals with one another and with the larger institution. Time spent in these efforts will be reflected in increased communication among units and identification of ways for increased synergy among units and the larger institution. Formal evaluation of division leaders must occur regularly and address the leader's effectiveness in facilitating goals of the individual faculty, the unit, and the institution. Recognition systems that reward collaboration and mentoring must be built into traditional promotion and reimbursement systems.

Emotional Intelligence

Most leadership manuals, psychology texts, and business leadership fables about organizational change concur that the starting place for healthy leadership is a substantial degree of self-awareness [7, 11]. Self-awareness is essential because the core values of the leader are reflected in the core elements of the culture. While it could be argued that a leader who has achieved significant success as a scholar or clinician surely has some degree of self-awareness, the issue for leadership is not efficient management or recognition of one's scientific orientation or values related to patient care.

Self-awareness in leaders pertains to a deep understanding of what motivates one to want to make a significant difference in the world. Such self-awareness entails knowing what motivates and drives one's values. A leader who is fixated on the bottom line because he/she was brought in to "clean this mess up" will create an authoritarian type of culture. By contrast, a leader who understands "no margin, no mission" and has a deep belief in the primacy of the patient and how it must drive everything in the organization, will create a very different, more collaborative organizational culture.

Self-awareness is part of a cluster of traits that have emerged as having major importance for leaders in academic medicine—emotional intelligence (EI). EI can simply be described as how we handle ourselves and our relationships and it has been identified as an essential dimension for leadership in academic medicine [21, 22]. A recent study of the leadership abilities of ten randomly selected chairs of departments of internal medicine concluded that "*emotional intelligence and its concomitant skills are the most essential competencies for leaders to succeed in academic institutions* [17]." Given the political and social stresses related to organizational complexity, fiscal constraints, diversity, and generational differences, it is easy to see why EI is essential for current leaders in academic medical centers. Leaders who wish to establish healthy, supportive environments for faculty must possess a solid foundation of EI, manage their own anxiety around issues such as promoting change, work well with teams, and manage conflict constructively.

Books and consultants proliferate to address self-management of leaders. Although specific strategies vary, agreement often exists about the utility of feedback mechanisms that allow leaders to gain some perspective on how others see them. Common strategies include 360 degree evaluations and workshops that highlight strengths and challenges of various individual work and personality styles [23]. Follow-up with small group work or individual coaching encourages leaders to assess their own fitness for leadership and implement changes in their style of interactions to achieve their desired results. Sometimes these programs are offered as rehabilitative plans. In contrast, a system that proactively embraces a culture of self-reflection and guided change in its leaders is more likely to be able to encourage the same self-reflection, self-regulation, and productivity of various kinds in faculty.

What if all leaders were skilled at such things as having difficult conversations with disruptive faculty? Would such emotional intelligence be sufficient to bring about systems change? It would not! The ability to manage a difficult conversation is an example of a skill that while absolutely necessary, is not sufficient for creating a healthy, aligned environment. Academic organizations function like any other system. The issue in bringing about behavior change in any system is not about a single intervention. Addressing ways to increase effectiveness is like any systems change and requires messages that are persistent, consistent, and congruent across all fronts

Difficult conversations are only one obvious need within healthy systems. Too often, "feedback" conversations only occur when there are changes to be made. Physicians and scientists expect to do well. The academic systems that "support" them also expect excellence, but paradoxically do not generally provide reinforcement of this success. Healthy systems create opportunities to celebrate achievement, identify what is working well, and amplify those positive successes. Popular organizational practices including "Positive Deviance" and "Appreciative Inquiry" are mechanisms to identify positive trends and stimulate what is working well [24, 25].

The Management of Trust

From developmental psychology, there is consensus that individual psychological health is fostered when the environment is safe [9, 10, 26]. Individuals within safe systems are free to engage in their own activities without worry that the larger system will collapse. The present uncertainty about many aspects of the health care system results in a general sense of scarcity and uncertainty that seeps through to faculty. Leaders who are trustworthy can help mediate this cycle of uncertainty and provide vision and a more secure environment in which faculty can better flourish.

Perhaps the best operational definition of trust is *consistency*. One may not like a leader's position with regard to a specific issue, but if the leader's position is consistent with his/her stated values and beliefs, he/she can be trusted. Trust may be the simplest, yet most effective, tool leaders have when dealing with complexity and change [20].

Leaders should not be expected to protect faculty from reality. They are also not expected to create a predictable environment when uncertainty reigns. However, they can be consistent themselves, in how they respond to emergencies and uncertainties, and in how they respond to faculty. Faculty should be able to trust that an institution's expectations for faculty are consistently applied, based on role and not favoritism. Faculty should be certain that a leader's word can be trusted. Building such trust requires leaders who are willing to make agreements thoughtfully and who are able to contain their own uncertainty to create a more trustworthy organization.

Incentive systems can be created that reward more than the generation of income or the publication of scholarly work. Division leaders must trust that their efforts to deal with problems or reduced effectiveness will be supported by senior leaders. Fair systems of post-tenure review, for example, can be instituted. Citizenship and mentoring can be clearly defined, measured, and reimbursed.

While Relative Value Unit (RVU) systems are frequently used to measure clinical contributions, some departments have expanded such systems to include contribution in teaching and service as well. One department has taken this to yet another level—the measurement of citizenship. Faculty are asked to rate each other on a general scale of characteristics that fall within the domain of "citizenship," characteristics such as supports and promotes colleagues, considers how one's behav-

ior affects others, does not abuse the rights of others, goes out of the way to help others, adheres to informal rules devised to maintain order, and attends meetings that are not mandatory but are considered important. Each faculty member then receives an averaged rating. This information is incorporated into the calculation of the department's bonus system. Reports from the faculty who participate in this system indicate that the overall working environment in the department has been positively affected through this effort.

Vision, emotional intelligence, and trust are all necessary elements of an aligned organization. Together these elements enable a leader to accomplish organizational goals, not at the expense of, but rather to the mutual benefit of faculty goals. Alignment of only certain aspects of the organization, such as clinical revenue generation, will not create organizational health. Alignment might best be compared to a cake recipe. Skipping certain ingredients because of the abundance of others will not result in a delicious or successful outcome. If health and success for faculty and the organization are desired, there can be no shortcut to aligning the organization's vision with the talent of its members, the resources available, and the rewards and incentives.

In the end, there is nothing magical about creating a psychologically healthy environment. Rather, faculty and leaders alike must accept responsibility for their shared contribution to the culture of the organization. Faculty must be willing to realistically assess their own talents and interests, and determine how they align with organizational goals and priorities. Faculty must also be willing to *speak truth to those in power* with regard to organizational inconsistencies. Leaders must have both the vision and the courage to align systems in a manner that ensures both individual and organizational success. Leaders must also be willing and able to listen about and see for themselves the inconsistencies that develop. Tables 5.1 and 5.2 offer a basic framework for assessing the psychological health of the individual and the institution of the academic medical center.

Table 5.1 Psychological health assessment of the individual faculty member

- Understands his/her attraction to a career in academic medicine
- Has clear personal and professional goals
- Understands how personal goals support organizational mission and goals
- Negotiates a role that is compatible with personal and professional goals
- Identifies an academic identity (area of excellence) or focus
- Identifies and contracts with mentors
- Develops an academic career plan
- Identifies areas for further development
- Is willing and able to be self-aware and to seek support and/or assistance as needed to maintain personal balance
- Is emotionally mature
- Accepts responsibility for one's own functioning within the institution

Table 5.2 Psychological health assessment of the Academic Health Center

- Has a clear mission and primary goals identified
- Communicates a clear set of expectations and responsibilities negotiated with each faculty member (alignment of individual and organizational goals)
- Provides mentoring and other systems to foster individual career success and assists in establishing personal–professional balance
- Aligns resources with organizational priorities
- Offers incentive and reward systems aligned with organizational priorities
- Conducts formative reviews (or feedback) with clear direction for how to get on track
- Provides substantive annual summative reviews
- Establishes organizational practices that identify and acknowledge faculty value and contributions
- Establishes programs and/or policies that afford latitude for faculty in dealing with changing life events or major career changes
- Implements strategies for helping individuals identify personal limitations and seek appropriate assistance
- Uses fair and systematic mechanisms to deal with disruptive behavior

Conclusion

Academic medicine has been described as both a major public trust and a national resource [19, 27]. But what is "academic medicine?" It is not clinics, laboratories, and classrooms. Academic medicine is the women and men who devote their lives to advancing medical science, developing the next generation of scientists, physicians, and other health professionals, and caring for the sickest within the society.

In a recent assessment of faculty vitality in a large Midwestern medical school, the two most highly ranked items that faculty considered important to vitality were having adequate time to accomplish goals and feeling valued for their contributions [28]. If faculty of academic medical centers can rightfully be seen as a critical resource, their well-being must be taken very seriously. Leaders do so by establishing a value-based culture and creating alignments throughout the organization that enable faculty to thrive and excel as scientists, educators, and clinicians.

References

1. Steiger, B. (2006) Doctors say morale is hurting *The Physician Executive* **32(6)**, 8–15.
2. United States Census Bureau (2006).
3. American Association of Medical Colleges (2006a) AAMC Graduation Questionnaire.
4. American Association of Medical Colleges (2006b) AAMC Data Book, Tables C4 and 5.
5. Schweitzer, L. (2007) The Status of PhDs in US Medical Schools, a plenary address at the Association of Psychologists in Academic Health Centers Meeting, Minneapolis, MN. May 3.
6. Mikesell, R., Lusterman, D.D., and McDaniel, S.H. (1995) *Integrating Family Therapy: Handbook of Family Psychology and Systems Theory* Washington, DC: APA Publications.

7. Bowen, M. (1978) *Family Therapy in Clinical Practice* New York: Jason Aronson.
8. Engel, G. (1977) The need for a new medical model: a challenge for biomedicine *Science* **196**, 129–136.
9. Bronfenbrenner, U. (1979) *The Ecology of Human Development* Cambridge, MA: Harvard University Press.
10. Bronfenbrenner, U. (2004) *Making Human Beings Human: Bioecological Perspectives on Human Development* (The SAGE Program on Applied Developmental Science) New York: Sage.
11. Friedman, E. (1985) *Generation to Generation* New York: Guilford.
12. Bickel, J., and Brown, A. (2005) Generation X: implications for faculty recruitment and development in academic health centers *Academic Medicine* **80**, 205–210.
13. Twenge, J. (2006) *Generation Me: Why Today's Young Americans are More Confident, Assertive, Entitled—and More Miserable Than Ever Before* New York: Free Press.
14. Schindler, B.A., Novack, D.H., Cohen, D.G., et al. (2006) The impact of the changing health care environment on the health and well-being of faculty at four medical schools *Academic Medicine* **81(1)**, 27–34.
15. Holloway, R., and Butler, D. (2007) Medical student distress and help-seeking behavior. 33rd Annual Predoctoral Education Conference of the Society of Teachers of Family Medicine. January.
16. Givens, J.L., and Tjia, J. (2002) Depressed medical students' use of mental health services and barriers to use *Academic Medicine* **77**, 918–921.
17. Lobas, J.G. (2006) Leadership in academic medicine: capabilities and conditions for organizational success *American Journal of Medicine* **119**, 617–621.
18. Institute of Medicine (2001) *Committee on Quality of Health Care in America. Crossing the Quality Chasm. A New Health System for the 21st Century* Washington, DC: National Academy Press.
19. Wilson, M.P., and McLaughlin, C.P. (1984) *Leadership and Management in Academic Medicine* San Francisco, CA: Jossey-Bass.
20. Bennis, W. (1984) The four competencies of leadership *Training and Development Journal* **38**, 15–19.
21. Goleman, D. (2002) *Primal Leadership: Realizing the Power of Emotional Intelligence* Boston, MA: Harvard Business School Press.
22. Grigsby, R.K., Hefner, D.S., Souba, W.W., and Kirch, D.G. (2004) The future-oriented department chair *Academic Medicine* **79**, 571–577.
23. Bachrach, D.J. (2002) The 360 assessment: using feedback from colleagues to improve performance. *Academic Physician and Scientist* **July/August**, 4–5.
24. Pascale, R.T., and Sternin, J. (2005) Your company's secret change agents. *Harvard Business Review* **May**, 1–10.
25. Whitney, D., Trosten-Bloom, A., and Cooperrider D. (2003) *The Power of Appreciative Inquiry: A Practical Guide to Positive Change* San Francisco, CA: Berrett-Kohler.
26. Erickson, E. (1968) *Identity: Youth and Crisis* New York: Norton.
27. Schroeder, S.A., Zones, J.S., and Showstack, J.A. (1989) Academic medicine as a public trust. *JAMA* **262**, 803–812.
28. Bogdewic, S. (2007) Faculty Development Needs Assessment. Unpublished survey.

Chapter 6
The Career Management Life Cycle: A Model for Supporting and Sustaining Faculty Vitality and Wellness

Thomas R. Viggiano and Henry W. Strobel

Abstract Over the past two decades, there has been much discussion in the medical literature about physician and scientist distress and little discussion about wellness. Only recently have academic health centers implemented dedicated office of Faculty Affairs to provide support to faculty. There are commonalities to some of the challenges faculty encounter in different phases of an academic career. In this report, we will describe a Life Cycle model for career management that may provide a framework to identify, understand, anticipate, and respond to the changing needs of faculty throughout their careers. The goal of this Career Management Life Cycle is to enable institutions to help each professional achieve and sustain professional vitality throughout their careers. The model may also help institutions collaboratively learn to effectively and efficiently use institutional resources to promote faculty well-being and assist faculty in health and illness.

Keywords Faculty wellness, faculty vitality, faculty life cycle, career management, career management life cycle, physician wellness, scientist wellness

Physicians and scientists who comprise the faculty of academic health centers have always had to overcome challenges to succeed in an academic career. During the past two decades, the challenges academic faculty confront have intensified. Diminished reimbursement for health care services has resulted in clinicians spending more time in clinical care activities and less time in academic pursuits [1]. There are also increased demands on educators to achieve and assess competencies in learners, yet education and clinical service activities are under-rewarded in academic

T.R. Viggiano
Associate Dean for Faculty Affairs, Office of Faculty Affairs, Department of Gastroenterology, Mayo Medical School, Rochester, Minnesota, USA
e-mail: viggiano.thomas@mayo.edu

H.W. Strobel
Department of Biochemistry and Molecular Biology, Office of Faculty Affairs, The University of Texas Medical School, Houston, Texas, USA

T.R. Cole et al. (eds.) *Faculty Health in Academic Medicine*,
© Humana Press, a part of Springer Science + Business Media, LLC 2009

promotions systems. Decreases in funding for research have made it more difficult for investigators to sustain their programs. Women and underrepresented minorities continue to be disadvantaged by less resource support, fewer mentors, and slower progression in academic rank [2].

Coincident with the increasing challenges to academic faculty over the past two decades, there has been more discussion in the medical literature about physician distress. Most of the discussion has focused on burnout, depression, anxiety, substance abuse, and difficulties in personal relationships. There has been little discussion about physician and scientist wellness and about how physicians and scientists cope with mental and physical illnesses [3]. Only recently have academic health centers implemented dedicated administrative structures to provide support to faculty [4]. As academic health centers establish these new administrative structures for faculty affairs, it is possible to design and incorporate support services that assist faculty during times of health and illness throughout their careers.

There are commonalities to some of the challenges faculty encounter in different phases of an academic career. If administrative professionals in academic health centers understand the needs of faculty in some of these predictable challenging professional experiences, they can design support services that anticipate and meet needs, and enhance the effectiveness and well-being of faculty. In this report, we will describe a life cycle model for career management that may provide a framework to identify, understand, and respond to the changing needs of faculty throughout their careers.

Career Management Life Cycle Model

The Career Management Life Cycle model describes experiences professionals may encounter throughout different phases of their careers. The model is grounded in the assumption that the relationship between an individual and an institution is formed for the purpose of meeting shared needs and realizing shared goals. The model is a developmental theory of professional development that is based on theories of individual development and organizational development. A developmental theory assumes there is a goal toward which an individual progresses. In the Career Management Life Cycle, the goal is to enable each professional to achieve and sustain professional vitality [5–7]. Vitality is a state in which there is optimal capability to realize shared individual and institutional goals [5–8].

There are eight phases in the Career Management Life Cycle: recruitment, orientation, exploration, engagement, development, vitality, transition, and retirement. Professionals who do not achieve or sustain vitality or who do not successfully manage transition may experience disengagement. Disengagement is an undesirable phase that compromises an individual's effectiveness, prevents them from experiencing the benefits of vitality, and may also be potentially detrimental to the professional's health or well-being. Disengagement may be costly to academic institutions and detrimental to the institution, patients, and society. Each phase of the Life Cycle presents needs and challenges to the individual, and an opportunity for the institution to provide meaningful support resources to assist the individual.

Recruitment

Recruitment involves all activities in which the individual and institution can assess and determine if there is potential for a long-term, mutually beneficial relationship. The best strategy for successful recruitment is to search for a "good fit" [6], that is, the alignment of the individual's and institution's values, needs, goals, and desired outcomes. Effective recruitment activities focus on the long-term goal of retention of productive individuals.

Recruitment should be viewed as an act of trust in which both the individual and the institution risk immersion in mutual commitments. The individual, department chairs, and potential mentors and collaborators should discuss the expectations and commitments involved in all work agreements. A document should be written that explicitly records work agreements so that all involved have a clear understanding of their reciprocal expectations [9]. Most important, the agreement should involve appropriate commitments from the individual and the institution and be both achievable and mutually beneficial.

Orientation

Orientation involves all organized activities that welcome and introduce the individual to the institution [5]. The goal of orientation is to provide acculturation and socialization experiences to the individual. Orientation provides an excellent opportunity to communicate important information to the individual and introduce them to colleagues who may provide guidance and support.

Effective orientation practices include both institutional and departmental efforts. Institutional efforts should celebrate the institution's heritage and culture, reaffirm the institution's mission and core values, and provide information about governance, policies, procedures, and support resources [10]. Departmental efforts should celebrate the department's history and accomplishments, and reaffirm the contract between the department and the individual. Departmental colleagues can facilitate connections within the institution and to the community outside the institution.

Exploration

Exploration involves all processes and activities that enable the individual to systematically investigate career options and opportunities. Faculty who are new to an institution or established faculty who are beginning a new endeavor may not be aware of all opportunities for professional development and collaboration within the institution. The goal of the exploration phase is for the individual to gain an understanding of the opportunities, expectations, support, and challenges for success in any chosen career path.

Engagement

The engagement phase involves all activities in which the institution helps the individual choose career paths and formulate career goals. Goal setting should be a collaborative process that involves the individual, mentors, and departmental leaders. We propose the mnemonic "SMARTER" to describe an effective framework for setting goals. SMARTER goals are specific, measurable, achievable, relevant, time-bound, equitable (in terms of benefit to the individual and institution), and respectful of the individual's personal life and commitments. Effective practices in goal setting include continual efforts to align the interests and needs of the individual and the institution, and plans to provide appropriate support resources and to monitor progress on achievement of shared goals.

Goal setting and performance reviews should be a continuum of activities. We also propose the use of an individual development plan [11] in which an individual writes a summary of their goals, activities, accomplishments, needs, and future plans to assist with performance reviews. The individual then summarizes progress on shared goals since the previous performance review and writes a self-assessment that includes a discussion of unmet goals, barriers to progress, needs for development, and revised goals with plans to address anticipated needs and barriers. The written individual plan document serves as a tool for communicating, planning, and assisting the individual to record, track, and assess progress on shared goals and to align support resources and development efforts to help the individual achieve those goals.

Development

Development includes any initiatives or activities that facilitate the ability of the individual to accomplish shared individual and institutional goals [12]. Development proceeds most effectively if there is continual alignment of eight "elements of development": goals, resources, training, mentoring, opportunities, assessments, rewards, and reflections. The individual development plan and periodic performance reviews are tools to maintain alignment of the "elements of development," and to guide and support the individual throughout their career.

The individual, mentors, and department leaders, should collaborate to maintain alignment of the elements of development. Effective performance reviews include not only discussion of the individual's professional goals and progress [13, 14], but also discussion of the individual's well-being and sources of stress. Stresses on the individual should be manageable and result in productive and non-counterproductive behaviors. Opportunities to enhance the individual's effectiveness should be identified and prioritized and written into the revised individual development plan. There should be mindfulness that the individual will only succeed if they are effective in balancing the challenges of both their personal and professional lives. Development is the responsibility of both the individual and the institution, and the ultimate measure of the success of development efforts is accomplishment of mission-aligned shared goals.

Vitality

Vitality is defined as the optimal capability of the individual to make significant and meaningful contributions to their career goals and the institution's missions [5–8]. Achieving and sustaining vitality throughout a career is the goal of the Career Management Life Cycle. The vitality of faculty is the lifeblood of the institution's professional community. Vital individuals execute the institution's missions, set the standards for professionalism and productivity, and are champions for the institution's values and cultures. Vitality is a synergistic state that cannot exist without mutual benefit of the individual and the institution.

The most difficult challenge to the institution is to help individual faculty sustain vitality over their entire career. Sustaining vitality is a challenging career-long journey for the individual because vitality is threatened by various destabilizing forces. As expertise is attained, one experiences fewer challenges, slower personal growth, and less motivation for some career goals [5]. Herzberg maintains that motivation from work itself comes from achievement, recognition, challenge, increased responsibility, advancement, and personal growth [15]. To sustain vitality, an individual must derive both motivation and fulfillment from their work.

The institutional environment can be conducive to sustaining the vitality of individual faculty members. Effective practice is to ensure that individuals have a manageable workload, equitable compensation, and sufficient support resources for scholarly work and professional development and renewal activities. Sustaining vitality is fostered by stimulating daily interactions with learners and colleagues, and an institutional environment committed to continual learning and improvement. An institution helps individuals sustain vitality by continually connecting people in relationships for mentoring and collaborative work, and by rewarding the most meritorious contributors with increased responsibilities and appropriate leadership positions. Most importantly, an institution's culture can help individuals sustain vitality by connecting individuals to the meaning and purpose of their work, expressing appreciation for each individual's contributions, and instilling a sense of community, or unity of purpose for accomplishing the institution's mission.

Transition

Transition describes periods in which faculty undergo significant changes in their career. Some transitions are evolutionary adjustments as individuals seek new challenges and responsibilities that result in continued personal growth and career enhancement. Other transitions involve redirection of faculty through major career changes (e.g., a surgeon who develops arthritis) or reengagement of faculty who have become disengaged.

Transitions can be very stressful, and faculty may benefit from appropriate assistance [5–9]. During evolutionary periods of transition, faculty may benefit from mentoring and career counseling. During major career transitions, individuals need to carefully evaluate their current situation and explore future options. Individuals may benefit from assistance to examine the disparities between their career aspirations and reality and identify the needs, issues, potential barriers, and support resources that affect each career option. An individual might benefit from assistance that helps them negotiate the appropriate balance of continuity and reinvention within their career. The individual should collaborate with department leaders and mentors to formulate new goals using the SMARTER method and to write a revised individual development plan.

Retirement

Retirement refers to the phase in which individuals prepare to withdraw from continuous active service to the institution. Retirement is not necessarily accompanied by a loss of vitality. Individuals may continue various levels of service to the institution and continue to contribute to the academic community as honored members.

One's professional identity contributes significantly to one's personal identity and institutions can prepare individuals for the altered identity of retirement [5, 6]. Institutions should help individuals explore opportunities for phased retirement and for appropriate meaningful and continuing involvement in the institution's mission. Retirees are repositories of wisdom and institutional history. They have lived through all phases of the life cycle and may serve as valuable mentors to assist active faculty through all phases of their career. Institutions can help individuals conclude their careers in a gratifying manner with celebrations and communications that convey that their life and work truly mattered.

Disengagement

Disengagement refers to the undesirable condition in which an individual's behavior or contributions are not aligned with the institution's values and mission. Disengagement is not a phase in the Career Management Life Cycle, but disengagement may occur

to any professional at any time. Examples of disengagement range from ethical or professional misconduct to persistently unacceptable productivity and misalignment of personal and institutional goals. The consequences of disengagement are significant. Persistent suboptimal productivity is not cost-effective and the individual and the institution are denied the rewards and benefits that are realized in the synergistic state of vitality. Furthermore, disengaged faculty can affect the morale of others, sometimes even having a toxic effect on the institution's culture.

Disengagement is not well described in the literature, and historically, institutions have treated disengaged faculty as individuals who have problems with attitude or motivation. Disengagement can be a very stressful period, and institutions often choose punitive rather than supportive measures with disengaged faculty [16]. Disengagement may occur when individuals have unmet or unhealthy desires for recognition, advancement, or influence. Disengagement may also result from alienation of individuals through real or perceived mistreatment by colleagues, leaders, or other institutional factors.

We believe that studying the reasons that faculty become disengaged will enable both individuals and institutions to better achieve and sustain vitality, reduce distress, and foster well-being. Some causes of disengagement are correctable, and institutions can learn how to minimize the probability and consequences of correctable disengagement and reengage faculty. Institutions may also benefit from recognizing uncorrectable causes of disengagement, e.g., ethical or professional misconduct, and implement and enforce policies that preserve our institution's values, culture, and resources.

Discussion

For over 20 years, there have been reports about burnout and distress among faculty in Academic Health Centers, yet little attention has been paid to their well-being. The prevalence of burnout of faculty has been estimated to be 37–47% and burnout has been associated with decreased career satisfaction [4]. There are also recent reports that burnout in physicians is associated with suboptimal patient care [4]. As Offices for Faculty Affairs emerge in Academic Health Centers, there is an opportunity to create structures that will be responsive to the needs of Faculty throughout their careers. We propose the Career Management Life Cycle model as a framework for understanding, anticipating, and responding to the needs of faculty and promoting their well-being.

The experiences and challenges of a professional's career unfold within a larger context of their personal lives. Faculty affairs professionals must have a heightened awareness and sensitivity to the impact of personal issues on professional activities. Professional support services may help individuals with their personal struggles [17], but faculty may be reluctant to seek help or not seek help. In the culture of an academic health center, individuals may avoid asking for help because they fear it could be viewed as an admission of weakness. The Career Management Life Cycle model may also provide a framework to guide the design of effective support services that anticipate and proactively respond to the needs of faculty throughout all phases of their career including times of health, sickness, and distress.

Comprehensive faculty support services include the triad of (1) professional development and career counseling, (2) personal and leadership development, and (3) services for personal and family matters. In addition to comprehensive health care, programs for child care, elder care, wellness and health promotion, personal development (which may include spouses), and basic leadership development should be offered. Institutions should provide ready access to confidential individual support services and public forums for ongoing open discussion of career matters [17]. Organizational development and leadership succession management benefit from programs that provide advanced leadership training to potential candidates for leadership positions.

The Career Management Life Cycle is a descriptive model of common and often predictable experiences that professionals may encounter in different phases of their career. The model may provide a useful framework for learning how to anticipate predictable faculty needs, prevent common problems, foster well-being, sustain professional vitality, and retain personnel. This descriptive model may become a prescriptive model that will enable institutions to share best practices and policies, and develop metrics to assess outcomes of career management programs, including the return on investment. The model may also help institutions collaboratively learn to effectively and efficiently use institutional resources to promote faculty well-being and assist faculty in health and illness throughout their entire careers.

References

1. Kuttner, R. (1999) Managed care and medical education *New England Journal of Medicine* **341**, 1092–1096.
2. Bickel, J., Wara, D., Atkinson, B.F., Cohen, L.S., Dunn, M., Hostler, S., et al., for the Association of American Medical Colleges Project Implementation Committee (2002) Increasing women's leadership in academic medicine: report of the AAMC Project Implementation Committee *Academic Medicine* **77**, 1043–1061.
3. Morahan, P.S., Gold, J.S., and Bickel, J. (2002) Status of faculty affairs and faculty development offices in US medical schools *Academic Medicine* **77**, 398–401.
4. Shanafelt, T.D., Sloan, J.A., and Haberman, T.M. (2003) The well-being of physicians *American Journal of Medicine* **114(6)**, 513–519.
5. Schuster, J.H., and Wheeler, D.W. (eds) (1990) *Enhancing Faculty Careers: Strategies for Development and Renewal* San Francisco: Jossey-Bass.
6. Baldwin, R.G. (ed.) (1985) Incentives for faculty vitality. In: *New Directions for Higher Education* No. **51**, San Francisco, CA: Jossey-Bass.
7. Bland C.J., and Bergquist, W.H. (1997) *The Vitality of Senior Faculty Members: Snow on the Roof, Fire in the Furnace* Washington, DC: George Washington University, School of Education and Human Development.
8. Clark, S.M., and Lewis, D.R. (eds) (1985) *Faculty Vitality and Institutional Productivity: Critical Perspectives for Higher Education* New York: Teachers College Press.
9. Menges, R.J. (ed.) (1999) *Faculty in New Jobs: A Guide to Settling in, Becoming Established, and Building Institutional Support* San Francisco, CA: Jossey-Bass.
10. Rice, R.E., Sorcinelli, M.D., and Austin, A.E. (2000) *Heeding New Voices: Academic Careers for a New Generation* Washington, DC: American Association for Higher Education.

11. Federation of American Societies of Experimental Biology. Individual development plan for postdoctoral fellows. Bethesda, MD [2002 (cited Jan 12, 2008)]. Available from: http:www. faseb.org/pdf/idp/pdf

12. Bland, C.J., Schmitz, C.C., Stritter, F.T., Henry, R.C., and Aluise, J.J. (eds) (1990) *Successful Faculty in Academic Medicine: Essential Skills and How to Acquire Them* New York: Springer, 22.

13. Braskamp, L.A. and Ory, J.C. (1994) *Assessing Faculty Work: Enhancing Individual and Institutional Performance* San Francisco, CA: Jossey-Bass.

14. Daniels, A.C. (1994) *Bringing out the Best in People: How to Apply the Astonishing Power of Positive Reinforcement* New York: McGraw-Hill.

15. Herzberg, F., Mausner, B., and Snyderman, B.B. (1959) *The Motivation to Work* 2nd ed. New York: Wiley.

16. Edwards, R. [1994 (cited Jun 25, 2004)] Toward constructive review of disengaged faculty *American Association of Higher Education Bulletin* [serial on the Internet] **48**, 6–7, 11–12, 16. Available from: http://www.AAHEBulletin.com

17. Hubbard, G.T., and Atkins, S.S. (1995) The professor as a person: the role of faculty well-being in faculty development *Innovative Higher Education* **20**, 117–128.

Chapter 7
Faculty Resilience and Career Development: Strategies for Strengthening Academic Medicine

Janet Bickel

Abstract This chapter outlines a number of promising directions for leaders in academic medical centers, who are forward-looking enough to place faculty vitality high on their list of priorities. Because the intersection of gender and generational issues is putting particularly complex demands on senior faculty, this chapter emphasizes both the work that remains to facilitate women realizing their potentials and the newer challenge of bridging generational differences.

Keywords Resilience, gender, generation, career development, mentoring

Introduction

Resilience is the capacity to remain robust under conditions of stress and change. Institutions that take a laissez faire attitude toward faculty resilience are putting their futures as academic institutions at risk [1]. This chapter outlines a number of promising directions for academic health center leaders forward-looking enough to place faculty vitality high on their list of priorities. Because the intersection of gender and generational issues is placing particularly complex demands on senior faculty, this chapter emphasizes both the work that remains to facilitate women realizing their potentials and the newer challenge of bridging generational differences. Recommendations for addressing this intersection include: (1) more flexible personnel structures; (2) updated approaches to mentoring and career development; and (3) greater support for department heads to improve faculty development practices.

J. Bickel
Career and Leadership Development Coach and Consultant, Falls Church, Virginia, USA
e-mail: janetbickel@cox.net

T.R. Cole et al. (eds.) *Faculty Health in Academic Medicine*,
© Humana Press, a part of Springer Science + Business Media, LLC 2009

Why Focus on Faculty Resilience?

During conditions of high stress and rapid change, professionals who can adapt will thrive. Those who cannot become less productive and less satisfied, and are more likely to burnout [2]. Resilience has been found to depend on: (1) remaining free of denial, arrogance, and nostalgia; (2) sound risk-taking and strategic experimenting with alternatives; (3) building one's community; (4) remaining guided by a high professional standards and one's core values; and (5) reflection and renewal. These indicators apply both to individuals and to organizations [3].

What does a resilient department look like? A primary characteristic is being forward-looking with respect to developing the talent it needs. In too many academic health center departments, the opposite holds true. Junior faculty are often treated more as expendable "fuel" for the clinical engine or as grant-writing machines than as junior members of an academic community and future leaders deserving of guidance and support.

Why should academic health centers be concerned about the resilience of its faculty?

(a) Baby Boomer faculty are graying and will soon begin retiring. Many departments already have long-standing unfilled vacancies. With many clinical departments resembling group practices more than academic communities and with the growing gap between salaries offered by private practice/industry compared to medical schools, many gifted physicians and scientists are opting out of academia. And over the course of their training, residents are becoming less rather than more interested in a faculty appointment [4]. With many schools under pressure to expand class sizes, will there be a sufficient supply of faculty, especially faculty who are academically productive and whom students view as excellent role models?

(b) Momentum in academic career development is not the force it used to be, and medical school faculty are less satisfied with their careers than in previous eras [5, 6]. Rising demands on faculty to see more patients and to generate more revenue and accelerating competition for grant funding are challenging many faculty beyond their capacities to succeed. Some have taken up residence in "Bitterness Valley," becoming vocally and visibly cynical.

A large Western region school of medicine recently measured the prevalence and determinants of the intent of its faculty to leave academic medicine. Over 40% were seriously considering leaving academic medicine in the next five years. Women and members of clinical departments were more likely to consider leaving than men or members of nonclinical departments [7]. A four-school study found that at least 20% of faculty had significant levels of depressive symptoms, with even higher levels in younger faculty. Over 20% of faculty reported thinking often of early retirement [8]. Clearly, depressed individuals or those intent on leaving are not likely to be great role models or very productive.

Academic health centers' lack of attention to its human resources is also expensive in terms of turnover. The costs of faculty turnover have been estimated to be

5% of Academic Health Center budgets (not including costs of lost opportunity, lost referrals, overload on other faculty, and reduced productivity and morale) [9]. While some turnover is predictable and healthy, too often it is the most rather than the least-valued individuals who leave.

(c) The educational programs of health care professionals and scientists do little to equip trainees to develop insight into their strengths, values, and motivations, or to take responsibility for their own career development. And faculty at many academic health centers find few resources or supports along these lines. Thus, many professionals lack critical skills in accurate self-assessment, understanding organizational cultures, adapting to change while maintaining focus, managing conflicts, building and participating in high-functioning teams, and communicating with people across differences. These faculty in turn are unable to fulfill their responsibilities as teachers and role models of these skills.

(d) While the traditional faculty path has assumed total immersion in the job, an increasing percentage of the incoming "intellectual capital" are women with substantial responsibilities at home. More young men as well are seeking opportunities that permit robust personal lives and ways to integrate family with professional life.

The Intersection of Gender and Generational Differences

Academic health centers are clearly dependent on the current trainees and junior faculty as their next high producers, mentors, and leaders. All of these individuals are, by definition, from a different generation and now over 60% of college graduates are women. Even three decades after women began entering the professions in force, despite their intellectual capital being equivalent to men's, they remain much less likely than men are to realize their professional potentials [10]. How can necessary improvements in biomedical research, medical education, and clinical care reach fruition if over half of the scientific and clinical intellectual capital remains underutilized?

Diversity is a necessary ingredient of an outstanding institution. Diverse teams outperform homogenous ones and in natural systems, as diversity increases, so does stability and resilience [11]. Yet, the paucity of women in board rooms, on key committees, and in leadership positions means that few men have experienced the improved collaborations, dynamics, and productivity that often results when a group is close to half men and half women [12].

The causes of gender disparities are mutlifactorial and remain controversial [13]. This chapter focuses on only two areas, both of which also affect many young men as well—the need for more effective mentoring and more flexible personnel policies.

While women faculty are as likely as men to report access to a mentoring relationship, most studies show that women gain less benefit from these relationships in terms of career planning and less encouragement to participate in professional activities outside the institution and that women are more likely than men to report that their mentor takes credit for their work and that their mentor is a negative role model [14]. Successful men who from boyhood have had role models reflecting their aspirations often take this advantage for granted and discount the extra challenges women face in building developmental relationships.

Some men are not as forthcoming or comfortable with women as with men mentees, which can impede the value of mentoring that women receive. Some men seem more comfortable in paternalistic relationships with women (i.e., father–daughter) than as equals. Then at the critical point when their women proteges begin spreading their wings and seeking more independence, it is not unusual for senior men to withdraw their support, as if put off by the protégé's growing power.

Being under-mentored seems to translate into a virtual "personal glass ceiling"—that is, women underestimating their own abilities and internalizing as personal deficits the cultural difficulties they face. These and other factors (e.g., being comparatively underpaid) reinforce each other, resulting in many women losing their ambitions and confidence and hence becoming less likely to successfully compete for raises, publications, and grants. Thus, senior faculty who make the extra effort to help women see their own leadership potential and connect them to resources and role models will likely have the satisfaction of witnessing a great impact on their development.

Lack of career guidance is not just a problem for women; virtually, all studies find that large percentages of faculty and residents of both sexes are not obtaining effective mentoring [15]. For example, one study found that only 42% were satisfied with the mentoring they received during residency, and 25% reported discomfort in discussing important issues with their mentor [16].

Many senior faculty are having difficulty adapting to their young proteges' different needs and realities [17]. It should go without saying the younger generations have had different life experiences and formative influences than their parents, but Baby Boomers tend to behave as if they were THE Generation and to forget how different the world is now from how it was in their youth.

One important difference is that Boomers have tended to define themselves through their jobs. Their children, Generation X, is the first one in which both parents were likely to work outside the home and in which parental divorce became prevalent. Corporate downsizing also took its toll during their youth. In part, because of these trends, Generation Xers are seeking a greater sense of family and have less faith in organizations than their Boomer parents did. They see the toll that such complete dedication has taken on their parents and mentors

Thus, among Generation Xers (most junior faculty and residents) and the Millennials (now emerging from medical and graduate schools), men as well as women are seeking more temporal flexibility than the typical faculty appointment allows. While committed to the profession, they are rejecting the pursuit of excellence through human sacrifice [18]. There is a gap between the model of success embodied by the career trajectories of men whose idea of balance is making it home in time for dinner to be served and what the younger generations are seeking (few of whom have nonworking partners). Current faculty who are successful by their discipline's and school's measures may not necessarily look like positive role models to Generation Xers and the Millennials.

Many of the younger generations of physicians and scientists measure success both by specific contributions to society and by their ability to maintain personal and professional balance. They do not want to disappoint their mentors, but they are explicitly creating and pursuing their own vision of this balance [19]. But instead of respect, too often young physicians encounter labels like "slacker" and value judgments about their commitment to the profession. As one resident asked, "Why are the older faculty so defensive and self-righteous—as if the way things were for them was the best of all possible worlds? If they really cared about us, they'd be trying to make life easier instead of hanging on to the past. Or maybe this is really about protecting their own privileges or justifying their own mistakes and sacrifices."

Promising Directions for Enhancing Faculty Resilience

Organizational, departmental, and faculty resilience depend on a supportive ecology such that individuals cooperate in achieving both institutional missions and in enabling each other to meet professional goals and potentials. In recent years, a "quarterly statement" orientation has replaced the longer-term perspective necessary to develop people and critical to building an organization's leadership "bench strength" and to succession planning. While there is an inherent polarity between short- and long-term perspectives [20], when a department resembles a giant adding machine more than a community of learners, then how successful can it be educating the next generation of scholars?

Academic health center leaders can better manage this difficult polarity with more attention to the following:

More Flexible Faculty Structures

Recruiting and retaining valuable faculty increasingly depends upon creating an environment in which individuals can build satisfying careers without having to choose between personal and professional success. But the continuing tyranny of the assumptions that it's "either advancement or family" and that faculty should not need time away from work especially during their twenties and thirties interfere with creative exploration of less-than-full-time alternatives.

Part-time pathways that can expand and contract as personal issues emerge are vital to making academe competitive with other medical career paths [21]. For example, part-time practice has been shown to be satisfying not only for physicians, but also for their patients [22]. In addition to nonpunitive less-than-full-time alternatives, other adaptive structures include opportunities to alternate high-involvement phases with lower involvement, unpaid leave for personal reasons without loss of benefits, and more off- and on-ramps for faculty as their responsibilities shift [23]. Other humane adaptations include post-service "catch up" time, mini-sabbaticals, bridge funding, and more backup for clinical services so one person's absence does not so negatively cascade onto others' backs.

While such options may incur some up-front costs, they are less expensive than re-recruiting and onboarding replacements. Offering temporal flexibility also translates into a competitive advantage, building commitment and loyalty in individuals who have many decades of professional life ahead of them. Moreover, in a field as demanding as medicine, why not try to support Gen-Xers' and the Millennials' intentions to integrate their personal and professional lives so that they achieve their potential?

Updated Approaches to Mentoring and Career Development

Burdened by increasing demands on them, faculty face many challenges in "being there" for the trainees and junior members of the academy looking to them for support and guidance. Mentoring represents the most tangible bridge to continuing traditions of excellence, but given how work-related demands have escalated and how informal time at the bedside has decreased, updated approaches to mentoring are necessary.

Mentoring entails effectively and comfortably communicating about many delicate issues, including beliefs about professionalism, and across many differences including gender and ethnicity. Patience is required, especially in establishing trust with someone very different from oneself. While administering oxytoxin to strangers has been found to facilitate trus, obviously no such shortcuts are available, and in any case nothing can substitute for listening with positive attention and asking open questions that encourage the other to explore their thoughts and feelings.

Faculty can be assisted to improve their competencies of active listening, avoiding assumptions, and combining an optimal balance of support and challenge, thereby maximizing their impact in the limited time available for this activity. Offering more learner-centered mentoring, they will avoid such common mistakes as undervaluing the younger generations' perspectives and automatically communicating their version of "reality." Fruitful opening questions are "What is most important to you right now?", "How do you most like to spend your time?", and "How would you define success in this situation?" If the answer to the first question is "finding competent child care," then begin with that.

Although some faculty seem irrevocably lost to a "hardening of the categories," most have something important to offer as mentors and with coaching can become more effective at and also get more out of this essential activity, extending their own legacies. Like senior citizens holding the paper further away so their eyes can better focus, mentors sometimes need to loosen their grip on how their proteges "ought to act." For example, rather than relying primarily on their greater expertise or beginning sentences with "When I was a resident," senior faculty need to remember and acknowledge how much higher the bar is now set in terms of the time pressures and the complexity of virtually every patient care decision.

Too often, mentoring is treated as an activity that faculty, who are so inclined, engage in during their "free time"; and there are no consequences for being a negative role model and mentor. But mentoring is a professional responsibility that medical schools should support and recognize as a core academic responsibility. Many senior individuals could use supportive coaching in acquiring the competencies entailed in mentoring "across differences," and this should be available.

The goal is building a supportive ecology in which collegial relationships develop as naturally as possible for all participants. In addition to one-on-one mentoring programs, an emerging model is collaborative and peer mentoring programs, for instance, facilitated group-mentoring that provides a framework for professional development, emotional support, and career planning [24, 25]. "Mentoring teams" can similarly be established so that junior faculty and trainees can access advice about an array of issues that a single mentor cannot adequately address [26]. This updating of mentoring practices also responds to medicine's need for new models of mutuality and "facilitative" leadership based on shared authority.

A well-staffed faculty development office should work with chairs in providing such opportunities. Such offices can also do more to help new faculty build their community and acclimate to the institution. However, few academic health centers currently offer comprehensive faculty development services or widely available leadership development programs [27].

Given that academic medicine is dependent on the current trainees as the next generation of faculty and leaders, academic health centers should also be actively encouraging trainees' interests in faculty roles. Residency programs should also offer trainees targeted assistance and coaching in seeking and negotiating the best possible first job; even if they do not stay within that academic health center or in academics, the trainees will appreciate and remember this assistance and support, which will build loyalty and help to recruit others. Residency and fellowship programs should also be offering trainees seminars on the whole range of career development skills, e.g., goal-setting, managing key relationships, and relationship-centered delegation.

Supporting and Evaluating Department Heads

Since department heads hold the keys to faculty vitality, faculty development needs to be high on their list of priorities. With the rising competition for funding and accelerating complexities in all knowledge and skill domains, it has never been a more challenging time to build a faculty career. Even though virtually all faculty are experiencing the same types of difficulties, each one feels alone with the steep challenges. So chairs should do whatever they can to create a sense of "academic community," including meeting with faculty regularly to provide constructive feedback and guidance regarding academic progress and professional development, and offering faculty opportunities for input into the department's governance. When people feel cared about and listened to, they are better able to rise to complex challenges and expectations.

Departmental leaders should be held accountable for their competencies as role models and as mentors, particularly in their ability to mentor "across differences." If junior faculty are given the opportunity to evaluate their chairs and mentors on such indicators as "provides timely feedback that both challenges and supports me," "advocates effectively for my development," and "inspires me as a role model," this will build a database that can be used for both summative and formative purposes.

During chairs' annual performance reviews, deans should also include assessment of the chair's progress in nurturing the careers of women and minority faculty. Linking effectiveness in these areas to a consequence of value, such as approval of new positions or access to faculty development resources, would add weight to this evaluation. Certainly department heads cannot mentor everyone, but they can create structures and options that assist all their faculty in obtaining the mentoring they need.

To accomplish these improvements, department chairs require support from their dean's offices. Excellent faculty affairs administrators build partnerships with chairs in becoming better developers of their human resources. They offer onboarding and educational sessions as needed and connect chairs who are struggling to coaches who can help them develop their "people" skills. A staple of leadership development in the corporate world, one-on-one executive coaching has been shown to increase the capabilities of motivated professionals particularly in the areas of accomplishing objectives and improving relationships [28].

Questions

To be sure, if these goals were easy to achieve, we would be further along. Many thorny questions arise as we consider how to boost faculty and departmental resilience. For instance:

1. How can academic health centers better support and assist faculty to take early responsibility for their own development and to see beyond their immediate career goals and departments such that they better understand their organizations and the "bigger picture" affecting them?
2. Even though faculty and leadership development programs likely protect investments and save resources in the long-run, how can strapped organizations come up with necessary start-up funding for such programs?

3. Where is the line between dedication to one's own family and health and insufficient dedication to one's profession and peers? How can we create safe forum in which to discuss such professionalism issues?
4. What academic health centers are making the most progress in aligning their performance evaluation criteria with their highest professional values and their educational missions (and not just their financial and research missions)? How can academic health centers better deal with administrators who use-up and demoralize faculty rather than develop them?

Conclusion

Assuring faculty resilience means addressing the developmental needs of current faculty and building bridges to the next generation. This work increasingly entails skillfully and courageously mentoring across many differences and creating alternatives to the traditional full-time model of continuous engagement. Unless academic medicine improves mentoring and expands its models, it will lose access to a great deal of intellectual capital. Neither academic medicine nor society can afford this loss.

References

1. Bland, C.J., Seaquist, E., Pacala, J.T., et al. (2002) One school's strategy to assess and improve the vitality of its faculty *Academic* Medicine **77**, 368–376.
2. Sotile, W., and Sotile, M. (2002) *The Resilient Physician: Effective Emotional Management for Doctors and Their Medical Organizations* Chicago, IL: AMA Press.
3. Westley, F., Zimmerman, B., and Patton, M. (2006) *Getting to Maybe: How the World Is Changed* New York: Random house.
4. Cain, J., Schulkin, J., Parisi, V., et al. (2001) Effects of perceptions and mentorship on pursuing a career in academic medicine in obstetrics and gynecology *Academic Medicine* **76**, 628–634.
5. Linzer, M., Konrad, T.P., and Douglas, J. (2000) Managed care, time pressure and physician job satisfaction: results from the physician worklife study. *Journal of General Internal Medicine* **15**, 441–450.
6. Anderson, W.A., Grayson, M., Newton, D., and Zoeller, E.D. (2003) Why do faculty leave academic medicine? *Journal of General Internal Medicine* **18(Suppl. 1)**, 99.
7. Lowenstein, S.R., Fernandez, G., and Crane, L.A. (2007) Medical school faculty discontent: prevalence and predictors of intent to leave *BMC Medical Education* **7**, 37.
8. Schindler, B.A., Novack, D.H., Cohen, D.G., et al. (2006) The impact of the changing health care environment on the health and well-being of faculty at four medical schools *Academic Medicine* **81**, 27–34.
9. Wenger, D. (2003) Conducting a cost-benefit analysis of faculty development programs *Academic Physician and Scientist* **May/June**.

10. Bickel, J., Wara, D., Atkinson, B.F., et al. (2002) Increasing women's leadership in academic medicine: report of the AAMC project implementation committee *Academic Medicine* **77**, 1043–1061.

11. Eagly, A., and Carli, L. (2007) Women and the labyrinth of leadership *Harvard Business Review* **September**, 63–71.

12. Bickel, J. (2007) The work that remains at the intersection of gender and career development. *Archives of Physical Medicine and Rehabilitation* **88**, 683–686.

13. Carr, P., Bickel, J., and Inui, T. (eds) (2004) *Taking Root in a Forest Clearing: A Resource Guide for Medical Faculty* Boston, MA: Boston University School of Medicine.

14. Pololi, L.H., and Knight, S. (2005) Mentoring faculty in academic medicine. *Journal of General Internal Medicine* **20**, 866–870.

15. Ramanan, R. et al. (2006) Mentoring matters: mentoring and career preparation in internal medicine residency training. *Journal of General Internal Medicine* **21**, 340–345.

16. Detsky, A.S., and Baerlocher, M.O. (2007) Academic mentoring—how to give it and how to get it *Journal of the American Medical Association* **297**, 2134–2136.

17. Bickel J., and Brown A. (2005) Generation X: implications for faculty recruitment and development in academic health centers. *Academic Medicine* **80**, 205–210.

18. Kalet, A., Fletcher, K., Ferdman, D.J., Bickell, N.A. (2006) Defining, navigating and negotiating success: the experiences of mid-career Robert Wood Johnson clinical scholar women. *Journal of General Internal Medicine* **21**, 920–925.

19. Johnson, B. (1992) *Polarity Management: Identifying and Managing Unsolvable Problems* Amherst, MA: HRD Press.

20. Froom, J., and Bickel. J. (1996) Medical school policies for part-time faculty committed to full professional effort *Academic Medicine* **71**, 91–96.

21. Parkerton, P.H. (2003) Effect of part-time practice on patient outcome. *Journal of General Internal Medicine* **18**, 717–724.

22. Socolar, R.S., Kelman, L.S., Lannon, C.M., et al. (2000) Institutional policies of U.S. medical schools regarding tenure, promotion, and benefits for part-time faculty *Academic Medicine* **75**, 846–849.

23. Pololi, L.H., Knight, S., Dennis, K., and Frankel, R.M. (2002) Helping medical school faculty realize their dreams: an innovative, collaborative mentoring program *Academic Medicine* **77**, 377–384.

24. Levy, B.D., Katz, J.T., Wolf, M.A., Sillman, J.S., Handin, R.I., and Dzau, V.J. (2004) An initiative in mentoring to promote residents' and faculty members' careers *Academic Medicine* **79**, 845–850.

25. Howell, L.P., Servis, G., and Bonham, A. (2005) Multigenerational challenges in academic medicine: UC Davis's responses *Academic Medicine* **80**, 527–532.

26. Morahan, P.S., Gold, J.S., and Bickel, J. (2002) Status of faculty affairs and faculty development offices in US medical schools *Academic Medicine* **77**, 398–401.

27. Fitzgerald, C., and Berger, J.G. (eds) (2002) *Executive Coaching: Practices and Perspectives* Mountain View, VA: Davies-Black.

Chapter 8
Diverse Academic Faculty: A Precious Resource for Innovative Institutions

Elise D. Cook and Harry R. Gibbs

Abstract Diverse academic faculty contribute unique perspectives and experiences that lead to creative growth of academic centers. Although the US population has become more diverse, academic faculty remain primarily heterosexual, able bodied, white, and male. These centers risk losing touch with the population at large and the issues they face. It is important to recruit and retain diverse academic faculty since they train future scientists and physicians who will make discoveries and apply treatments to the entire population. There is a paucity of data about diverse academic faculty and their unique additional stressors impacting on faculty health. In this chapter we discuss these stressors as they apply to race and ethnicity and faculty with disabilities. We also examine the important associations between marginalization, isolation, and silence experienced by diverse faculty and the stress that follows.

Keywords Disability, faculty, faculty health, faculty with disabilities, minority faculty health, diverse faculty health, underrepresented minority, faculty of color

Many stressors have been identified affecting academic faculty health, from the rank of instructor to tenured professor and administrator. When compared with their majority counterparts, diverse academic faculty members have higher levels of occupational stress due to additional unique stressors. In this chapter we will describe some of these additional stressors and a few successful programs implemented in academic centers as solutions. Racial and ethnic minority populations are defined as American Indian and Alaska Native, Asian, Black or African American, Hispanic or Latino, and Native Hawaiian, and Other Pacific Islander. The Association

E.D. Cook

Associate Professor, Department of Clinical Cancer Prevention, The University of Texas M. D. Anderson Cancer Center (M. D. Anderson), Houston, Texas, USA.
e-mail: edcook@mdanderson.org

H. R. Gibbs
Vice President, Office of Institutional Diversity, M. D. Anderson, Houston, Texas, USA

T.R. Cole et al. (eds.) *Faculty Health in Academic Medicine*,
© Humana Press, a part of Springer Science + Business Media, LLC 2009

of American Medical Colleges (AAMC) classifies US Citizens or Permanent Residents who indicate Black, American Indian/Alaska Native, Native Hawaiian or Mexican American, and Puerto Rican Mainland ethnicity in combination or alone as underrepresented minorities (URM). Non-URM includes all other minorities and "unknown" race [1]. The diverse groups presented in this chapter include underrepresented minority (URM) and faculty with disabilities. Unique challenges faced by women are covered in Chapter 7. Although many unique stressors exist for other diverse groups such as gay, lesbian, bisexual and transgender, foreign-born, and non-underrepresented minority faculty, these groups are not discussed in this chapter.

America Needs Diverse Academic Faculty

As the US population becomes increasingly diverse, URM clinical and research faculty ensure that the health care system is representative of the nation's population and responsive to its health care needs. For example, in the 2000 US Census data, Blacks, Hispanics/Latinos, and Native Americans comprised nearly 26% of the US population. However, in 2004, Blacks, Hispanics/Latinos, and Native Americans comprised only 6.4% of all physicians, who graduated from US allopathic medical schools. Even more disturbing, the number of URM health care providers is decreasing [2].

Increasing diversity in the health care workforce is a potential solution to health care disparities because studies have shown that patients prefer physicians of their own cultural background, medically indigent patients are more likely to be served by minority physicians, and minority physicians tend to locate their practices in racial/ethnic minority communities thereby providing access [3–8]. URM faculty members play an important role by training these medical and science professionals to meet the needs of our increasingly diverse US population.

The scarcity of URM basic science faculty is even more alarming. According to Mervis, the number of Blacks hired as assistant professors at the nation's top 50 chemistry departments has held steady—at zero—between 1991 and 2001. One faculty stated "A lot choose industry not for the money, but because the workplace is cut and dried. The bottom line is clear, and hard work is rewarded." Minority faculty chemists say that the status quo is maintained by practices such as failure to interview qualified minority candidates, loading down a new faculty with introductory courses, and marginalizing senior scientists. For example, one faculty member who was assigned a heavy teaching burden speculated that these classes might have been a well-meaning attempt to attract minority undergraduates to the department. He stated, "Before I got tenure, I taught a freshman course every year. I found out later that no other young faculty, before or since, had done that. … It put me at a big disadvantage when I came up for tenure." Very little information exists about the unique stressors affecting other URM non-MD research scientists such as population, behavioral health, public health scientists, and epidemiologists [9].

Medical Faculty

There is little literature available that addresses the challenges faced by medical faculty with physical and mental disabilities. Certainly, these individuals are faced with many of the same issues that surround any group that is "different" from what society feels is "normal". However, the dimension of diversity that includes people with disabilities is unique in a number of ways:

1. Disabilities may be invisible. A person's gender, race, and relative age are frequently apparent to anyone who can observe the individual. Even socioeconomic status can be estimated by one's dress, mode of transportation, size of home, etc. Unfortunately, a disability may not be evident to the naked eye. Heart disease, emphysema, arthritis, mild hearing deficits, learning disorders, etc. may all be present to the degree that they affect one's performance, but are not readily apparent.

2. Disabilities can be acquired. Again, most primary dimensions of diversity are present from birth, and cannot be changed. Disabilities can strike anyone at any age at any moment. Thus, many faculty with disabilities find themselves suddenly transferred from being members of the majority to being members of a minority.

3. Disabilities can affect performance. A person's gender, race, religion, sexual preference, etc. usually has no impact on his/her ability to perform most tasks. However, physical and mental disabilities can have an impact upon one's abilities to adequately perform the expected duties of a medical practitioner (without appropriate accommodation).

When looking at these issues, it may be helpful to analyze them according to three sets of challenges that faculty with disabilities must meet: Access, Accommodation, and Attitude.

Access

Individuals who have been disabled since birth or early childhood travel a different path than those who acquired a disability as an adult. On one hand, they have had a lifetime to learn coping skills in order to survive and prosper in a world that, for the most part, assumes that everyone is healthy. On the other hand, they have had to find unique ways in order to get the education required to become a physician or scientist. Although the national census reports that one in five individuals has a disability which significantly affects his/her life, it is estimated that less than 1% of medical school graduates are people with disabilities [10]. These numbers suggest that it may be especially difficult to become a physician if one has a significant disability. Accommodation problems (especially in older academic institutions, where expensive accommodations may be difficult to accomplish, and in nonprofit institutions, where resources for accommodations may be scarce) certainly exist. However, there is also controversy in some centers as to what physical skills are required for a successful career in medicine. Many institutions feel that medical training programs should be undifferentiated. In other words, every medical student should be able to become a psychiatrist, orthopedist, pediatrician, etc., and, if you cannot perform the duties that would be required for one specialty, you cannot be a successful physician. Recently, scholars have suggested that a more differentiated approach, in which students choose paths earlier in their training, would result in less-rigid curricula, and provide more opportunities for students whose skills in one area may not be as well developed as in others [11].

Faculty with acquired disabilities have not had to face these challenges in the past. Thus, the fact that they do not have easy access to certain facilities, licenses, opportunities for career change and growth, etc., comes as an unwelcome shock. The coping skills that others learned in their growth and development have to be learned anew. They have to not only deal with the physical issues surrounding their illness or condition (including medication, physical therapy, and speech therapy), but they also have to face the reality that their entire life now has to undergo a major adjustment. The fact that this may occur at crucial moments in their career, such as during the tenure clock, or while being considered for promotion or leadership assignment, adds a large amount of stress to an already stressful occupation.

Accommodation

Title I of the Americans with Disabilities Act requires employers to provide reasonable accommodations that do not cause them "undue hardship". These accommodations include physical access, work schedules, training procedures, and reassignment to other jobs [12]. While most medical facilities understand these provisions when it comes to patients, there are fewer acknowledgements of these standards when it comes to accommodating faculty. In their recent study of medical institutions, Steinberg and associates [13] reported numerous stories of difficulties that faculty with disabilities have faced in terms of accommodation, from having to enter buildings through loading docks, to being unable to get to various classrooms and conference rooms because of a lack of elevators. Adequate signage and technical aids to hearing are frequently not available. Not all restrooms may accommodate individuals with disabilities, which means longer time spent during meeting breaks to attend to biological necessities. At our own institution, automatic doors are quite common in patient areas, but are much less evident in those parts of the medical center which are frequently traveled by faculty and students. Lack of adequate plans for evacuation of employees with disabilities in case of emergencies also leads to a feeling of helplessness.

Accommodation issues are particularly problematic for those faculty with newly acquired disabilities, or with a disability that did not require special accommodations in the past, but has now progressed over time. Frequently, there is reluctance for departments to recognize that the individual was able to perform without accommodations before, but now requires extra equipment, job assignment changes, etc.

Attitude

Steinberg reported that physicians with disabilities have mixed experiences in terms of being accepted by their peers. Pity, compassion, jealousy, fear, and anger are frequently reported emotions. There is a feeling of "unfairness" reported, as accommodations which are made for those with disabilities are viewed as an unfair advantage not afforded to the rest of the faculty. A common theme that occurs in these reports is that most institutions are only interested in the compliance aspects of dealing with disabilities, rather than the more humanistic one of providing faculty with the tools they need in order to succeed in their careers. In addition to the normal stress that all faculty face, including tenure clocks, publish or perish attitudes, increasing patient loads, and diminishing external resources for research, faculty with disabilities frequently feel as if they have to overcompensate to try and remove doubts that others may have about their ability to perform at a level equal to their peers. While this feeling has also been reported by women and minorities in the medical profession, faculty with disabilities frequently are faced with physical barriers that are not present for these other groups.

Faculty who were "healthy" in the past, and now ask for accommodations because of a newly acquired or progressing disability, are sometimes viewed as "slacking off", or even faking their disabilities to obtain compensation or "special treatment". This is especially evident if the disability is an illness or condition that is not readily apparent. In addition, there is the "pity factor", which frequently leads to a reduction in responsibilities, job change, or refusal to promote because of a feeling that it would be "too much for the individual to handle", even if this is not the case.

All members of marginalized groups report feelings of oppression, loneliness, and alienation. These feelings are certainly prevalent in groups of faculty with disabilities. However, the uniqueness of the actual physical limitations to success, as well as the fact that many of these individuals did not face these challenges in the earlier portions of their careers, can have a magnifying effect upon these emotions. Interestingly, medical faculty with disabilities may be more accepted by medical students, especially in a teaching role, than those who are fully "abled" [14].

Stress-Inducing Factors for Underrepresented Minority Faculty

Collegiality

Marginalization of diverse faculty members is a primary source of occupational stress and creates a sense of being invisible, social isolation, and devaluation of one's work in the academic community [15]. Stanley reported on the experiences of faculty of color teaching in predominantly white colleges and universities. She evaluated narratives of these experiences to break the silence. One contributor writes about his experience with invisibility: "One of my colleagues was honored with a university-wide award. Embarrassingly, only I and one other member of my department attended the event. The next day, the dean promptly called the department chairperson, chastising the department for their absence while praising, by name, the *one* faculty member that was present. The only person of color at this event, sitting right in front, next to the colleague now publicly named, I was present but not seen, noticed perhaps, but not remembered" [16]. Minority physicians also report concerns about being invisible to their colleagues if they are not wearing their white coats [17].

One contributing author writes about her positive experience with mentoring across differences: "Even where the number of Latinos or faculty of color is slim, seek out a diverse network of committed teachers. They not only provide you with an extra set of eyes and ears for the classroom; they can also provide you with the type of honest feedback both you and your students require to succeed. While senior white faculty cannot address all the dilemmas encountered by Latinos, many have successfully navigated troublesome classroom waters. You owe it to yourself to avail yourself of their considerable knowledge and experience" [16].

Another author writes how a mentor facilitated her transition from a nonacademic position to the professoriate: "My work at the National Research Council (NRC) enhanced my professional development. It was at the NRC that I developed a research interest in human resource issues in the science and engineering workforce focusing on underrepresented groups—African Americans, Hispanics, and non-Hispanic white women. The executive director of my NRC unit actively encouraged my scholarly activities—including publishing. He understood academy and did everything he could to facilitate my return to the classroom. He was extraordinary. … Actively seek out mentors whether or not they are assigned" [16].

Price et al. reported structural barriers (poor retention efforts, lack of mentorship, and cultural homogeneity) that hinder URM faculty's success and professional satisfaction after recruitment. Other factors that hindered URM faculty's success were limited networking opportunities. For example, one focus group male participant stated "I think you almost need a critical mass to have an effective network, and unless you have that critical mass of people, there's no minority-specific networking possible." Another URM female participant stated: "With numbers come comfort, and if you're 'going to' be the only one there, I think it's 'going to' be much more uncomfortable than if you're joining a group of, you know, ten or fifteen, even if it's in a large sea of people" [17].

Thomas performed a three-year study on the career progression of minorities in major US corporations and found that whites and minorities follow two distinct patterns of advancement. Although this is a study of corporate professionals, some of the findings of his study may apply to academic faculty. Specifically, promising white professionals tend to enter a fast track early in their careers, whereas high-potential minorities take off much later. He also found that the people of color who advance the furthest share one characteristic, a strong network of mentors and corporate sponsors who nurture their professional development [18]. According to Thomas, "Mentors must be willing to give their mentees the benefit of the doubt: they invest in their mentees because they expect them to succeed. But a potential mentor who holds negative stereotypes about an individual, perhaps based on race, might withhold that support until the prospective mentee has proven herself worthy of investment. (Such subtle racism may help explain why none of the minority professionals in my study had been fast-tracked. Whites were placed on the fast track based on their perceived potential, whereas people of color had to display a proven and sustained record of solid performance—in effect, they often had to be over prepared—before they were placed on the executive track.)" [18].

Thomas reported that the networks of minority executives were also much more diverse than those of the minority managers who did not attain executive status. He found that having all mentors of the same race or all white mentors were not as beneficial as building genuine, long-term relationships with both whites and members of same race.

Thomas also found that when white mentors understand and acknowledge race as a potential barrier, they can help their mentees deal effectively with some of those obstacles. This open discussion of racial issues yields greater opportunity for the mentee [18].

Lower Promotion Rates

Palepu, Peterson, and Fang reported that URM faculty members are less likely to be promoted when compared with white faculty; this finding was not explained by potential confounders such as years as faculty or measures of academic productivity [19, 20]. URM faculty have more clinical responsibilities, less research time, more need to supplement their incomes, are less satisfied with their careers, and are more likely to leave academic medicine within five years [19, 21]. Fang reported that when compared with white associate professors, URM faculty were less likely to be promoted regardless of tenure status or receipt of National Institutes of Health (NIH) grant funding. In general, minority faculty were more likely to be affiliated with departments and medical schools with lower promotion rates, and are less likely to be on tenure tracks, or to receive NIH awards, the two strongest positive predictors of promotion. Asian Pacific Islanders and URM faculty were more likely to be women, for whom lower promotion rates have been well documented. Even after controlling for the percentage of time spent in clinical work, African American faculty members were still less likely to hold senior rank. Minority faculty members often make contributions in less-valued promotion criteria such as, education, administration, and community service [19]. All minority faculty members may be discomforted by lower rates of promotion and concerned by the knowledge that they may face many barriers to advancement.

Racial/Ethnic Discrimination in Academic Medicine

Racial and ethnic discrimination have contributed to the lack of URM faculty in academic medicine and especially in senior leadership positions [19, 21]. Peterson reported that URM faculty and non-URM faculty also had significantly more odds than majority faculty of experiencing racial/ethnic bias in their professional advancement. Faculty members with such experiences had lower career satisfaction scores than other faculty ($P < .01$) and were less likely to feel welcome at their institution. Despite this additional stress, they received comparable salaries, published comparable numbers of papers, and were similarly likely to have attained senior rank (full or associate professor) [22]. Although these data may reflect that minority faculty are able to overcome these experiences, these data may also reflect that the researchers did not capture the experience of all minority faculty. There was a 40% nonresponse rate to the survey and the experience of minority faculty who had already left academic medicine could not be captured. This study examined racial/ethnic discrimination by superiors and colleagues only and did not explore other possible sources of such problems, including patients, hospital staff, and students [22]. Cora-Bramble postulates that the cohort of minority faculty in Peterson's study may have been able to thrive because of the theoretic construct of resilience. The construct of resilience includes intrinsic assets such as self-efficacy, self-esteem, ability to deal with stress, cultural identity, and communication and negotiation skills, as well as external resources such as mentorship, faculty development, and minority faculty peer groups. Further research into resilience-based solutions could expand our understanding of minority faculty retention and advancement in environments toxic to both [23].

Having a primary language other than English was associated with the experience of racial/ethnic bias, independent of minority status. According to Price "Foreign-born faculty noted that having a foreign accent is a frequent source of scrutiny. ... 'Lack of language fluency' may negatively impact important career advancement opportunities for foreign-born faculty" [17]. Faculty without access to Standard American English may face linguistic scrutiny affecting all levels of communication, including presentations and interactions with colleagues and patients. African American faculty members who speak with the African American dialect rather than the Standard American dialect often face linguistic scrutiny. John Baugh coined the term linguistic profiling, a new name for an old form of racial discrimination that occurs after hearing the sound of someone's voice over the phone. The caller is identified as either Hispanic or African American and the apartment, job, or loan is "no longer available." African American dialect was formed over many generations by descendants of African slaves. Their forefathers were deprived of their native language on slave ships through systematic separation, denied the ability to read or write, and future generations had limited exposure to Standard English [24]. Linguistic discrimination is not going away any time soon, so some authors recommend learning Standard American English as a second language rather attempting to replace the African American dialect [24]. This embarrassing, demeaning form of discrimination is similar to the preconceived notions some observers have that people who speak the Queen's English are smarter and more important or those who speak with a southern dialect are not very bright. This form of discrimination must be exposed for all to see its ridiculousness and the pain that it causes.

Carr reported that minority faculty members frequently perceive academic medicine discrimination as a battle, bombarded on many levels. A fundamental part of this discrimination is described as being ignored or excluded, in academic circles and at social functions. One URM respondent to a study conducted by telephone interview stated: "Scientifically, it is difficult to thrive if you are isolated. ... You have to look for avenues ... to interact with colleagues on a professional level." Another stated: "I don't think it is color as much as it is who you know. Because the majority has a broader network ... one gets invited to present a major talk because you know somebody" [25]. However, many instances of racial discrimination are perceived as resulting from ignorance in that they stem from a general lack of knowledge or absence of exposure to minorities rather than from deliberate discrimination. One URM faculty stated, "I don't think people mean to do it; it is something they do unconsciously. ... Whether it is done in ignorance or intentionally, it ... has the same results" [25]. Carr stated that minority women may find it difficult to determine whether offensive, harassing, or discriminating behavior is gender-based or ethnicity-based [26]. In a focus group, one participant summarized her experience of being female and URM: "[I]n terms of your everyday work, patient care and interactions, that may be a little difficult and is worse in the beginning ... as people get to know you, they become a little less gender conscious and color conscious. And they then start feeling you more as a colleague. But in the beginning, it's very hard, very hard" [27].

Another challenge that minorities face is deciding when to confront bias and stereotypes in the workplace without negatively affecting their career development. This indecision about confronting racial and ethnic bias in the workplace is a significant source of stress. One URM male responded during a telephone interview study: "Fish or cut bait—either don't do anything or go all out, there is no in-between." An URM female in the same study responded: "As minorities you reach certain levels, you don't want to lose what you have ... keep quiet about it ... you may lose more by fighting it out." However, most faculty in this telephone interview study felt it was important to acknowledge and confront discrimination. For example, a URM female stated: "Sit down with colleagues ... make your case, and raise the level of awareness." For a number of interviewees, not addressing these situations was not an option: "Minority faculty need to be proactive ... if you don't tend to it, nobody deals with it." Another stated: "Most people who are doing these things are people you can talk to ... have a discussion ... find out what the situation is ... if it is not to one's satisfaction, then go up the ladder." 'Going above' in the medical hierarchy until satisfaction was achieved was recurrently suggested [25].

Baez studied faculty members resisting perceived racism in the promotion and tenure process. Some faculty who "fought back" but failed, believed they acted to effect significant change. One Asian American faculty member perceived being "labeled hostile" for bringing up issues of race and gender in the promotion and tenure process. She failed to obtain tenure. Three faculty members were afraid of getting too angry or overly sensitive, did nothing about perceived racism, appeared somewhat demoralized and felt that even if they did not get tenure, they were learning lessons about surviving the promotion and tenure process at another institution.

Some faculty members did not see "fighting back" as completely detrimental to the attainment of tenure, learned to "pick and choose" their battles by striking a balance between not "putting people off" and "staying true." These faculty members decided early in their careers which things were "worth fighting for" and which were "not worth the trouble." Other faculty resisted racism by forming or joining networks of other faculty of color who provided professional and interpersonal support, while others treated the promotion and tenure process as a game and believed that if they learned to play the game they would survive. However, one of the most important ways that faculty resisted institutional racism was to redefine the promotion and tenure criteria in ways that benefited their racial communities and special interests of faculty of color [28].

Community and Academic Service

Diverse academic faculty members are often asked to fulfill socially responsible roles that may take time, but not lead to academic advancement. According to Stanley, many minority faculty hesitate to serve the university and community as diversity experts, they do so because they know that if they do not, the diversity voice gets lost at the table. These activities benefit the institution, but research in these areas is considered to have "less rigor" and when these same come up for promotion or tenure, these activities are not given serious weight in the process [29]. In fear of failure to achieve promotion and tenure, some faculty refrain from participating in community service activities, an honored tradition especially in African American, Native American, and Latino communities. An African American saying, "Remember where you came from" is an example of community service expectations; whether or not these obligations are honored, additional faculty stress ensues. The African proverb, "She who learns must also teach," echoes the importance that historically has been attached to black faculty women's efforts to share their knowledge with others regardless of any opposition or challenges they may have faced [30]. Mentoring students and residents is a service activity viewed as a way to give back to the community. Interestingly, even though URM faculty mentor students more often, URM faculty members are less likely to have mentors.

Among Gregory's 1999 study sample of 384 black faculty women nationwide, the primary work activity of most of those employed at four-year colleges and universities was teaching, followed by administration and research. All of her respondents typically engaged in more teaching, advised greater numbers of students, and participated in more committee work than did their white male counterparts. They subsequently conducted less research and published fewer articles than did either white faculty men or women. They further tended not to be included in collaborative research projects with their peers. Many minority faculty members whose work has been published in journals that focus on minority issues have reported that their majority peers sometimes fail to recognize the quality of their research, and focus instead on where they published [30]. For black women faculty, this can create a complex challenge because many tend to focus their research in the very areas where they teach and provide community service [30].

These data are very concerning since URM family medicine faculty were 2.7 times more likely to have moderate to severe depression and these odds increased with increasing time devoted to teaching and increasing stress scores [31].

Racial and Ethnic Bias in the Classroom, Lab, and Clinic or Hospital

Many faculty of color report experiences in as well as outside of the classroom that include challenges to their authority and expertise, negative behaviors and attitudes of students, and complaints being made to senior faculty and administrators about their teaching [29]. In a study of minority faculty in predominantly white universities, one of the contributors provided an example of how the male students sometimes would try to show that she did not know her material. She stated, "after I had explained a point in class, a male student would attempt to explain the point again in a manner that suggested my explanation was incorrect. From the tone of the student and the timing of the comment, I felt he was trying to demonstrate that I, this black woman professor, was not knowledgeable." Another contributor, an URM associate professor in health and kinesiology stated, "Some of my students have challenged my credentials in and outside of the classroom and have had the audacity to question my appointment at a 'superior' institution of higher learning. … I have actually had a parent tell me, "My husband is an M.D., and he read his daughter's paper, and she should have received an A. Are you sure you know how to write and teach?" I wonder. … If I were a white male tenured faculty member, would I have been approached like this?" [29].

Recommendations to Institutions

Opportunities for Promotion and Tenure

Institutions should determine if disparate promotion and tenure exist in their institutions. If so, evaluate promotion and tenure criteria in terms of how much emphasis is placed on research compared with community service, education, and administration. Promotion criteria should be evaluated to ensure that teaching, clinical, and administrative activities are sufficiently rewarded. If URM faculty are asked to serve on institutional committees, such efforts should be accounted for and steps should be taken to appropriately adjust their teaching, clinical, and research loads for this participation. Also, administrators should work to nominate, support, and mentor faculty of color for leadership development opportunities and positions [16, 19].

Faculty Development Programs

Provide a faculty development program to increase retention and provide the educational and socialization experiences necessary to reduce gender and racial disparities in academic promotion [25]. If a program does not exist, consider developing one at your institution, sending minority faculty members to programs at other institutions, or sending faculty to the annual AAMC Minority Faculty Career Development Seminar [32].

University of Texas M. D. Anderson Cancer Center Faculty Development Programs

Although M. D. Anderson has no faculty development program dedicated to URM faculty, several faculty development programs exist and are available for all junior and senior faculty:

1. The Faculty Leadership Academy (FLA) is a one-year leadership development program designed specifically for faculty leaders at M. D. Anderson. Content covers leadership skills, diversity issues, increasing ones' own self-awareness as a leader, and improving team effectiveness. It facilitates multidisciplinary collaborations within departments and throughout the institution, and develops confidence in taking charge of a complex academic department or team. Senior leadership nominate faculty for FLA participation.
2. The Faculty Leadership Graduate Seminars are held two to three times annually and are planned so that FLA graduates can delve deeper into issues they find most challenging.
3. The Heart of Leadership: Core Skills course, offered to all new and current assistant professors, provides an opportunity to gain a greater understanding of the M. D. Anderson culture. The program is a two-day condensed version of the Faculty Leadership Academy (FLA), designed specifically for faculty not yet in formal leadership positions. The interactive sessions allow new and current assistant professors to build the interpersonal skills needed to effectively manage and lead staff and to collaborate successfully in teams across the institution.
4. The Junior Faculty Development Program will hold its second retreat for incoming faculty that introduces them to their peers and to mid- and senior-level faculty, as well as provides information and resources necessary to successfully navigating the M. D. Anderson environment.

Creighton Minority Mentoring Program

1. Creighton University School of Medicine (CU SOM) has a pipeline mentoring program targeting students from the fourth grade through graduate school. CU SOM also has separate programs mentoring women and minority faculty

members with the singular purpose of providing junior faculty members of underrepresented groups with one or more mentors who can provide them with the information, connections, and support necessary to succeed in their academic careers. All URM faculty in the CU SOM Center of Excellence (COE) program meet at least once with the director of faculty development for the COE; the URM faculty provide input to assist pairing faculty with mentors. If possible, senior faculty members with similar backgrounds and racial/ethnic origins are selected as well as mentors with more specific academic connections. The department chair is always included as one of the mentors unless delegated to another department faculty member. Preliminary data after the first 18 months since implementation of the COE program revealed that all URM faculty members in the program have remained at Creighton for one year or longer. The five-year retention rate for the first year of the mentoring program was 58%, as opposed to 20% prior to implementation of the program. Only one URM faculty has resigned in the past two years, while six new URM faculty have been recruited, bringing the total number of full-time URM faculty and administrators in the SOM in 2005 to 23 (7.5% of the total faculty, up from 6.9% before implementation of the program) [33].

2. The proportion of tenured/tenure-track individuals has risen from 25% to 44% of URM faculty over the past three years, with more nontenure-track individuals planning to transfer to the tenure track over the next year. These preliminary data suggest that even in its early stages, the COE Faculty Mentoring and Faculty Development Program at CU SOM is successfully assisting faculty in their career advancement [33].

UCSD and NCLAM Mentoring Program

Using information obtained from the informal UCSD Hispanic Center of Excellence faculty interviews, the UCSD SOM established the National Center for Leadership in Academic Medicine (NCLAM). The UCSD NCLAM is a structured mentorship program that addresses the professional development needs of junior faculty by providing the knowledge, skills, and resources necessary to have successful careers in academic medicine. Junior faculty received counseling in career and research objectives; assistance with academic file preparation; introduction to the institutional culture; workshops on pedagogy and grant writing; and instrumental, proactive mentoring by senior faculty. For NCLAM participants, the retention rate of URM junior faculty at UCSD and in academic medicine (87% and 93%, respectively) is nearly equal to the retention rate of all other junior faculty at UCSD and in academic medicine (82% and 95%, respectively) [34]. Their confidence in skills needed for academic success improved: 53% personal leadership, 19% research, 33% teaching, and 76% administration. Given improved retention rates, the savings in recruitment was greater than the cost of the program [35].

Address the Institution's Diversity Climate

Addressing an institution's diversity climate requires: (1) an institutional culture that encourages faculty to openly express their opinions and insights, (2) an institutional culture that makes faculty feel valued, and (3) the institution to incorporate faculty perspectives into its main mission and culture [18].

M. D. Anderson's Office of Institutional Diversity (IOD) participates in recruitment activities to enhance the institution's efforts to recruit and retain a diverse workforce and provides diversity educational programs for the institution's faculty and staff. The OID also sponsors the Employee Networks (Organization for Women, Organization for Minority Employees and Gay, Lesbian, Bi-sexual and Transgender Network).

Diversity Accountability Data

Administrators should develop data management systems for accountability in terms of faculty and diversity. Data on date of hire, rank, tenure-track status, race and ethnicity, retention, attrition, and exit interviews could help to determine the rate of progress and identify barriers to institutional growth and development [16].

Recommendations to URM Faculty

Participants in the study by Carr et al. provided these recommendations to faculty dealing with discrimination[25]:

(a) Repeatedly, minority faculty described the need to be strongly self-reliant and to be realistic. Minority faculty stated the need to excel as well as to be continually tested and to work harder than most to earn a place at the table.

(b) Minority faculty believed they should educate and lead by example; exposure to minority excellence is seen as a way to help resolve misunderstandings of cultural differences and overcome preconceived ideas. This can evolve to socializing once ignorance is overcome, and socializing can lead to ever-improving understanding and discourse. Recurrently, the strategy for managing racism was perceived to involve education, as well as creativity and ability.

(c) Minority faculty recounted the need to have adequate support to succeed in academic medicine, including noninstitutional support from sources such as family, clergy, and community: Support in personal and professional arenas was needed to navigate, as well as a broad range of skills and training in negotiation, conflict resolution, management, research techniques, and writing grant applications.

Table 8.1 Response strategies for racial and ethnic discrimination

1. Take time before reacting to an episode of discrimination to get distance and perspective
2. Discuss the situation with a trusted colleague or mentor
3. Gather information regarding the parties involved, the department, and institution using deliberate and thoughtful fact finding
4. Be clear about the objectives you are seeking and be sure they are reasonable
5. Confront the discrimination in a thoughtful and levelheaded manner without hostility
6. Consider informing colleagues about the issue so that it can be a learning experience for majority faculty (managing up)

(d) Effective networking was also important to provide opportunities: being included in the right network was perceived as key to obtaining opportunity and as possibly more important than race.
(e) When discrimination was encountered, minority faculty overwhelmingly described the need to remain calm, maintain a clear perspective (Table 8.1), and not respond immediately.
(f) Diversity (sensitivity) training was viewed as a means of improving the climate in academe and increasing the awareness of majority colleagues to the challenges that minority faculty face.

Recommendations to Enhance Diversity and Improve URM Faculty Health

Boyer suggested that the work of the intellectual life included scholarship not only of discovering knowledge (research), but also of integrating knowledge, of applying knowledge, and of teaching. After Boyer's death, Glassick provided the standards for evaluating this full range of scholarly work. He stated: "Knowledge for all the glory and splendor of the act of pure discovery, remains incomplete without the insights of those who can show how best to integrate and apply it." Through application of these principles and guidelines, URM and all faculty members would benefit from a system that rewards the expertise represented by faculty work [36].

URM faculty may use the methods described by Meyerson for "tempered radicals" to succeed in academic medicine. She describes "tempered radicals" as people who want to succeed in their organizations yet want to live by their values or identities, even if they are somehow at odds with the dominant culture of their organization. She provides mechanisms for "tempered radicals" who want to fit in, retain what makes them different, and use their differences to inspire positive changes in their organizations. She recommends maintaining positive self-definitions by building relationships with people, who share these marginalized aspects of your identity, manage heated emotions, and create an appearance of conformity, while acting on differences to sustain your sense of self. She also recommends that when dealing with personal threats, "tempered radicals" recognize that a response to any particular interaction is a personal choice. Difficult interactions can be viewed as opportunities, view silence as sometimes the best choice, consider the complex "self" in that threats may be aimed at only one aspect of self to help evaluate other aspects of the situation rather than launching an all-out attack to defend the one "true self" and depersonalize encounters [37].

Conclusion

According to Cohen, "A racially and ethnically diverse faculty, fully empowered by the equitable presence of minorities within all ranks of the academy, is the only conceivable bridge to the diverse physician workforce and the culturally competent health care system that the full spectrum of the American public deserves. As long as our medical school faculties have little more than token representation from many sectors of the richly diverse American culture, and as long as faculty advancement, for whatever reason, is grossly distorted by race and ethnicity, the medical profession cannot truly lay claim to the ethical and moral high ground it professes to occupy" [38].

URM faculty in predominantly white academic centers often manage chronic daily racism and other stressors, URM women faculty have more unique stressors, URM women faculty with disabilities have even more unique stressors, and so on. These individuals also have unique perspectives and bring different abilities to the academic table. Presently, we do not have enough diversity in our pool of scientists and physicians to serve our diverse population needs. Diverse faculty members will teach future physicians and scientists to serve the health care needs of our diverse country; they will attract other diverse faculty and students and allow this process to continue.

Academic centers have a long way to go to address the issues concerning recruitment and retention of URM faculty. In many institutions, the promotion and tenure process is unbalanced, new methods of evaluating and rewarding meaningful scholarship have been devised, and application is all that is needed in most cases. However, resistance may be encountered. Changing the reward systems in academic centers will cause some of the privileged majority to lose; however, overall there is much more to gain.

We will recover lost productivity from URM faculty suffering from anxiety and stress while (1) trying to maneuver a system that does not work for them, (2) working harder than most to earn a place at the table while sacrificing a fulfilling personal life, (3) providing meaningful community service or conducting community-based participatory research—also known as marginal work, knowing or not knowing (due to lack of mentorship) that performing this work will put them at a disadvantage during the promotion and tenure process, (4) showing up for meetings only to be "invisible" to others, (5) not being invited to socializing events, to present at prestigious meetings or to collaborate on research projects, or (6) experiencing the pain of isolation.

URM faculty must do their part as well by reaching out to colleagues. Ask them to lend their experience by reviewing a manuscript or grant application; ask to attend an event that most of the other members of your section are attending; meet, greet, and ask questions at meetings to remove the cloak of invisibility. Read, review, and seek to understand your institution's requirements for promotion and tenure to guide your future academic commitments. Seek out faculty development programs at your institution, at other institutions, or the AAMC Minority Faculty Career Development Seminar.

Suggested recommendations for dealing with the issue of accommodating faculty with disabilities are listed in Steinberg's article. Simply providing these faculty with appropriate accommodations would go a long way to alleviating the stress of trying to pursue a medical career with a disabling condition. In addition, the medical community needs to understand that these individuals have a unique lens with which to view the landscape of medical care and research, and can perform to the same standards as anyone else as long as they are provided with reasonable accommodations. Those who feel that this is an unnecessary burden would do well by counting the number of medical personnel who wear glasses at their next staff meeting. Similar principles apply to scientific researchers.

Faculty and administrators at academic centers are some of the brightest individuals around. By embracing the differences in our colleagues, vital and empowering differences will be accomplished in academic medical centers.

References

1. FACTS—Applicants, Matriculants and Graduates—Applicants, Accepted Applicants, and Matriculants by URM Status and Gender, 2002, http://www.aamc.org/data/facts/archive/famg52002a.htm (Accessed January 12, 2008).
2. Diversity in the Physician Workforce: Facts & Figures 2006, Association of American Colleges, *Academic Psychiatry* 30, 3.
3. Moy, E., and Bartman, B.A. (1995) Physician race and care of minority and medically indigent patients *Journal of the American Medical Association* **273**, 1515–1520.
4. Komaromy, M., Grumbach, K., Drake, M., et al. (1996) The role of black and Hispanic physicians in providing health care for underserved populations *New England Journal of Medicine* **334**, 1305–1310.
5. Brotherton, S.E., Stoddard, J.J., and Tang, S.S. (2000) Minority and nonminority pediatricians' care of minority and poor children *Archives of Pediatric Adolescent Medicine* **154**, 912–917.
6. Stinson, M.H., and Thurston, N.K. (2002) Racial matching among African-American and Hispanic physicians and patients *Journal of Human Resources* **37**, 410–428.
7. Cooper-Patrick, L., Gallo, J.J., Gonzales, J.J., et al. (1999) Race, gender and partnership in the patient-physician relationship *Journal of the American Medical Association* **282**, 583–589.
8. The Diversity Research Forum, Exploring Diversity in the Physician Workforce: Benefits, Challenges, and Future Directions, Convened by the Division of Diversity Policy and Programs at the AAMC'S 2006 Annual Meeting, Seattle, WA.
9. Mervis, J. (2001) Minority faculty: new data in chemistry show 'Zero' diversity *Science* **292**, 1291–1292.
10. Wu, S.S., Tsang, P., and Wainapel, S.F. (1996) Physical disability among American medical students *American Journal of Physical Medicine and Rehabilitation* **75**, 183–187.

11. DeLisa, J.A., and Thomas, P. (2005) Physicians with disabilities and the physician workforce: a need to reassess our policies *American Journal of Physical Medicine and Rehabilitation* **84**, 5–11.
12. West, J. (1991) *The Americans with Disabilities Act: From Policy to Practice* New York: Milbank Memorial Fund.
13. Steinberg, A.G., Lezzoni, L.I., Conill, A.,and Stineman, M. (2002) Reasonable accommodations for medical faculty with disabilities *Journal of the American Medical Association* **288**, 3147–3154.
14. Pfeiffer, D., and Kassaye, W.W. (1991) Student evaluations and faculty members with a disability *Disability and Society* **6**, 247–251.
15. Thomas, G.D., and Hollenshead, C. (2001) Resisting from the margins: the coping strategies of black women and other women of color faculty members at a research university *The Journal of Negro Education* **70**, 166–175.
16. Stanley, C.A. (2006) Coloring the academic landscape: faculty of color breaking the silence in predominantly white colleges and universities *American Educational Research Journal* **43**, 701–736.
17. Price, E.G., Gozu, A., Kern, D.E., Powe, N.R., Wand, G.S., Golden, S., and Cooper, L.A. (2005) The role of cultural diversity climate in recruitment, promotion, and retention of faculty in academic medicine *Journal of General Internal Medicine* **20**, 565–571.
18. Thomas, D.A. (2001) The truth about mentoring minorities: race matters *Harvard Business Review* **79**, 98–107.
19. Palepu, A., Carr, P.L., Friedman, R.H., et al. (1998) Minority faculty and academia rank in medicine *Journal of the American Medical Association* **281**, 767–771.
20. Fang, D., Moy, E., Colburn, L., and Hurley, J. (2000) Racial and ethnic disparities in faculty promotion in academic medicine *Journal of the American Medical Association* **284**, 1085–1092.
21. Palepu, A., Carr, P.L., et al. (2000) Specialty choices, compensation and career satisfaction of underrepresented minority faculty in academic medicine *Academic Medicine* **75**, 157–160.
22. Peterson, N.B., Friedman, R.H., Ash, A.S., Franco, S., and Carr, P.L. (2004). Faculty self-reported experience with racial and ethnic discrimination in academic medicine *Journal of General Internal Medicine* **2004**, 259–265.
23. Cora-Bramble, D. (2006) Minority faculty recruitment, retention and advancement: applications of a resilience-based theoretical framework *Journal of Health Care for the Poor and Underserved* **17**, 251–255.
24. Rice, P. (2006) Linguistic profiling: the sound of your voice may determine if you get that apartment or not http://news-info.wustl.edu/tips/page/normal/6500.html (Accessed January 13, 2008).
25. Carr, P.L., Palepu, A., Szalacha, L., Caswell, C., and Inui, T. (2007) Flying below the radar: a qualitative study of minority experience and management of discrimination in academic medicine *Medical Education*, **41**, 601–609.
26. Carr, P.L. (2000) Faculty perceptions of gender discrimination and sexual harassment in academic medicine *Annals of Internal Medicine* **132**, 889–896.
27. Waitzkin, H., Yager, J., Parker, T., and Duran, B. (2006) Mentoring partnerships for minority faculty and graduate students in mental health services research *Academic Psychiatry* **30**, 205–217.
28. Baez, B. (1998) Negotiating and Resisting Racism: How Faculty of Color Construct Promotion and Tenure, Georgia State University. Research Report, 1998, http://search.ebscohost.com/login.aspx?direct = true&db = eric&AN = ED430420&site = ehost-live (Accessed January 13, 2008).
29. Smith, M.A., Barry, H.C., Dunn, R.A., et al. (2006) Breaking through the glass ceiling: a survey of promotion rates of graduates of a primary care faculty development fellowship program *Family Medicine* **38**, 505–510.
30. Gregory, S.T. (2001) Black faculty women in the academy: history, status and future *Journal of Negro Education* **70**, 124–138.

31. Costa, A.J., Labuda Schrop, S., McCord, G., and Ritter, C. (2005) Depression in family medicine faculty *Family Medicine* **37(4)**, 271–275.
32. Hamilton, K. (2005) Changing the face of American medicine *Diverse: Issues in Higher Education* **22**, 32–34.
33. Kosoko-Lasaki, O., Sonnino, R.E., and Voytko, M.L. (2006) Mentoring for women and underrepresented minority faculty and students: experience at two institutions of higher education *Journal of the National Medical Association* **98**, 1449–1459.
34. Daley, S., Wingard, D.L., and Reznik, V. (2006) Improving the retention of underrepresented minority faculty in academic medicine *Journal of the American Medical Association* **98**, 1435–1440.
35. Wingard, D.L., Garman, K.A., and Reznik, V. (2004) Faculty success, outcomes and cost benefit of the UCSD national center of leadership in academic medicine *Academic Medicine* **79**, S9–S11.
36. Glassick, C.E., Huber, M.T., and Maeroff, G.I. (1997) *G.I. Scholarship Assessed: Evaluation of the Professoriate, An Ernest L Boyer Project of the Carnegie Foundation for the Advanced of Teaching*, San Francisco, CA: Jossey-Bass, 1–11.
37. Meyerson, D.E. (2003) *Tempered Radicals: How Everyday Leaders Inspire Change at Work* Boston, MA: Harvard Business School Press, xi–xii; 41; 59–60.
38. Cohen, J.J. (1998) Time to shatter the glass ceiling for minority faculty [Editorial] *Journal of the American Medical Association* **280**, 821–822.

Part IV
Perspectives from the Humanities and Interpretive Social Science

Chapter 9
Organizational Culture and Its Consequences*

R. Kevin Grigsby

Abstract Typically comprised of a medical school, teaching hospital, and one or more physician practice groups, the complex organizational structure of academic health centers can be difficult to understand, especially to persons who are unfamiliar with academic medicine. Adding to the complexity, each academic health center has a unique organizational culture. Most broadly defined, "culture" is a shared pattern of basic assumptions shared by a social group about itself. Organizational culture describes these patterns of basic assumptions and related behavior within a defined social environment. Organizational culture has a major effect on faculty health. Some academic health centers may be described as having a "conflict laden" culture, where competition and the associated struggle to remain "on top" are a way of life.

Some academic health center faculty members refer to an environment where compensation is based on a model where "You eat what you kill." On the other hand, some organizational cultures are more nurturing and, as such, more conducive to maintaining the health of the faculty. "Family-friendly" cultures may offer generous leave to families following the birth or adoption of a child. Likewise, a high priority may be placed on mentoring of junior faculty. In recent times, preventive health and wellness programs have become a part of the organizational culture of some academic health centers. Nonetheless, others persist in the expectation of long work hours, subordination of personal needs like sleep and meals, and self-denial in general. Other changes in organizational culture could enhance the health and well-being of faculty at academic health centers.

R.K. Grigsby
Professor of Psychiatry, Department of Psychiatry, Vice Dean for Faculty and Administrative Affairs, Penn State College of Medicine, Hershey, Pennsylvania, USA
e-mail: kgrigsby@psu.edu

*Portions of this chapter were presented at the *Faculty Health Conference* sponsored by The John P. McGovern Center for Health, Humanities, and the Human Spirit at the University of Texas Health Science Center, Houston, Texas, July 19–21, 2007.

Keywords Academic health centers, organizational culture, faculty health and well-being

Introduction

Edgar Schein argues that leaders create culture and that an understanding of culture is necessary if leaders really intend to lead [1]. As such, leadership and organizational culture are inextricably linked. So, too, is the effect of leadership and organizational culture on the health of faculty members in the academic health center. This chapter offers information on the relationship between organizational culture, leadership, and the consequences of organizational culture related to the health of faculty members in the academic health center.

Organizational Culture

Typically comprised of a medical school, teaching hospital, and one or more physician practice groups, academic health centers are complex. Some academic health centers include other professional schools including public health, dentistry, nursing, and allied health. Understanding the organizational structure is daunting, especially to persons unfamiliar with academic medicine. Adding to the complexity, each academic health center has a unique *organizational culture*. Most broadly defined, "culture" *is a shared pattern of basic assumptions shared by a social group about itself.* In an oft cited, succinct, and easily understood definition of *culture*, Bower defines it as "the way we do things around here" [2]. In defining the culture of the workplace, Peterson and Wilson explain that basic assumptions "form an unspoken or unwritten basis upon which people behave, communicate, and interact in the workplace" [3]. *Organizational culture* describes these patterns of basic assumptions and related behavior within a defined social environment [3].

Organizational Culture in Academic Health Centers

Although there are many similarities between academic health centers, especially in terms of organizational structure, it is probably most accurate to say "If you've seen one academic health center, you've seen one academic health center" as there are often are vast differences when comparing the *organizational culture* of one academic health center with another. As in any large social system, the enduring nature of organizational culture is not conducive to change. Patterns of behavior become engrained and those persons entering the organization are "swept away" by the prevailing currents. Unless specific actions are taken, nothing will change. If a leader or anyone else in an organization wants the culture to remain the same, she or he does not have to do a thing. The consequences of taking no action have the same net result as making a conscious decision to keep things from changing. Likewise, the negative features embedded in the culture will endure and continue to have a deleterious effect. The "way we do things" within the culture of the academic health center affects the health and well-being of the faculty members comprising the organization. As such, aspects of organizational culture affect faculty health and well-being.

Aspects of Organizational Culture Leading to Decreased Faculty Well-Being

Some academic health centers may be described as having a "conflict laden" culture, where competition and the associated struggle to remain "on top" are a way of life. Some academic health center faculty members refer to their workplace environment as a place where compensation is based on the "You eat what you kill" model, explaining that the environment is highly competitive and that rewards are commensurate with an aggressive approach to performance. While some individuals may thrive in such an environment, organizations operating under this credo can create an environment where other individuals experience a high degree of *organizational pain*. Too often, this is manifested in a loss of self-worth, feelings of hopelessness, and a loss of energy and drive for some of the individuals comprising the organization [4]. Individuals manifesting these symptoms are not limited to faculty members—these effects may be felt by everyone within the organizational culture. For example, Dahlin and Runeson report that third year medical students often report mental distress and burnout. Recognizing that attributing these effects to individual psychological profiles alone is faulty, they suggest that "organizational as well as individual interventions may be needed" [5].

The human experience can be characterized across physical (body), affective (heart—as in the vital center of one's emotions), cognitive (mind), and spiritual (soul) dimensions. When organizational culture detracts from, as opposed to enhancing, any or all of these dimensions, the net result is decreased health and well-being. While all "work" requires investment and, at times, self-sacrifice, inordinate amounts of physical, emotional, mental, and/or spiritual stress detracts from faculty health and well-being. Prolonged periods of self-sacrifice, lack of recognition for a job well done, overt physical or emotional abuse, and the presence of "burned out" faculty and staff define an organizational culture as healthy or unhealthy—and members comprising the organizations reflect the culture in the relative presence—or absence—of health [6].

Aspects of Organizational Culture Leading to Enhanced Faculty Well-Being

On the other hand, the organizational cultures of some academic health centers are more nurturing and, as such, more conducive to maintaining or enhancing the health of the faculty.[1] "Family-friendly" cultures may offer generous leave to families following the birth or adoption of a child. Likewise, a high priority may be placed on mentoring of junior faculty [7]. Some academic health centers have begun to create an organizational culture where enhancing personal fulfillment of faculty physicians is recognized as critical to the success of the organization [8]. Preventive health and wellness programs have become integral parts of the organizational culture. Nonetheless, the organizational cultures at other academic health centers persist in the expectation of long work hours, subordination of personal needs like sleep and meals, and self-denial in general. Across the nation, the implementation of the "80 hour work week" represents an organized initiative to improve the quality of work-life for residents in training in academic health centers. On the other hand, it is unclear that the same level of commitment to limiting work hours is in place for academic health center faculty. When resident work hours were limited, the organizational culture of academic health centers had to change. At some academic health centers, physician faculty members were expected to absorb some of the impact of limited hours by working longer—and harder—to compensate for the loss of resident hours. On the other hand, other academic health centers had resources to allowing them to develop alternatives to asking for "more and harder" effort of faculty members. For example, at some academic health centers, a consequence of the 80 hour work week has been the addition of mid-level providers to offset the workload of house staff and attending physicians. There are many opportunities for other changes in organizational culture that could enhance the health and well-being of faculty members, as well as resident physicians, at academic health centers.

Leadership and Organizational Culture

Organizations begin to create cultures "through the actions of the founders who operate as strong leaders" [1]. Even in long-established organizations, the influence of early leaders is often manifested in the present-day actions of members. Fundamentally, the enduring nature of organizational culture makes it especially daunting for the new leader who desires to make changes in the organizational culture. A new leader may enter an organizational culture and find

---

[1] For example, Indiana University School of Medicine and the *Relationship-Centered Care Initiative*; *The Next Generation* plan at Penn State College of Medicine and Milton S. Hershey Medical Center; and the *Engaging Excellence* initiative at SUNY Upstate Medical University.

it to be defensive, counterproductive, and dysfunctional for the organization as a whole" [9]. In fact, the leaders may find a "culture of stress" and recognize the need for change. Creating a healthy workplace requires more than cleanliness, attractive surroundings, and safety. In fact, there are many other components of culture to be considered as the leader promotes a healthy workplace culture. For example, consideration of gender differences may lead to greater motivation, increased productivity, increased organizational loyalty, and improved physical and mental well-being if concomitant changes are made to better meet the needs of women in the organization [10]. Such practices would include stopping the tenure clock during the key child rearing years or allowing nursing mothers to schedule time to nurse infants. Locating child care facilities in close proximity to the mothers' worksite can reduce the burden of travel on the mother and promotes interaction with their children.

Increasing diversity of the workforce, students, and patients requires active promotion of accepting more, and different, persons where in order to achieve a good "cultural fit" [11]. Creating organizational policies and practices that are responsive to "difference" whether it pertains to age, disability, race, ethnicity, sex, socioeconomic status, national origin, or gender orientation is critical if organizations expect to change as the demography of our nation changes. For example, providing space to allow Muslim personnel to participate in daily prayers, creating a gay, lesbian, bisexual, and transgender support group that meets on site, or acknowledging and sponsoring a celebration of Juneteenth allows organizations to promote the well-being of all personnel, rather than limiting recognition and celebration to the holidays and traditions of the majority culture.[2] Leaders should understand that addressing culture may be a key factor in improving workplace health by reducing or eliminating workplace stress experienced by the faculty. On the other hand, ignoring workplace stress leads to undesirable outcomes for both individuals and the organization. Creating a culture within the academic health center that enhances the professional fulfillment of physicians will include attention to personal factors such as the nature of the work, recognition for a job well done, and growth as part of overall change initiatives [8]. Leaders must consider both individual (interpersonal relationships) and organizational (workplace conditions) factors simultaneously in order to affect health [12]. Workplace stress is one important factor to consider if the health of faculty is to be enhanced. Low job satisfaction and higher perceived stress among physicians appear to be related to "greater intentions to quit, decreased work hours, change of specialty, or leave direct patient care" [13].

---

[2]Juneteenth, also known as Freedom Day or Emancipation Day, is the oldest known holiday celebrating the end of slavery in the United States. It originated in 1865 and is celebrated on June 19 every year.

Wilfred Drath argues that leadership is a property of an organizational culture; no individual has sole possession of leadership. Leadership is what happens when individuals within the organizational culture collaborate in thoughts and action. In fact, followers often exert more influence over formal leaders than those leaders exert over individuals in the organization [14]. Nonetheless, formal leaders can set a course for change, commit resources to change, and initiate collaboration in thoughts and demonstrated action. However, the tasks of promoting and improving faculty health require far more than the vision of a single leader. The fundamental challenge to leaders is to successfully engage *everyone* in the organization to intentionally change organizational culture so that faculty health and well-being are not only promoted, but are integral to the culture itself. A fundamental question for leaders should be: How do I engage everyone comprising the organization to change the ways we do things around here?

Changing "The Way We Do Things Around Here"

Changing the organizational culture of an academic health center may require "reinventing" the academic health center. Even in the best situation, changing organizational culture is daunting. In one attempt to make significant changes at the Penn State College of Medicine and Milton S. Hershey Medical Center, the process was initiated by completing a campus-wide cultural assessment [15]. Gaining an understanding of the organizational culture and the "culture code or codes" within the organization guides leaders to act on accurate information decisively, rather than to act in a hit or miss fashion. A culture code is a single word or phrase that captures the values, assumptions, and behaviors that define the culture within an organization [16]. Understanding culture codes should not be limited to those leaders "at the top," but should be understood by leaders at several levels of the organization. The specific role of leadership in managing change occurs "through different mechanisms at different developmental stages of an organization" [1, 3–11]. Educating a cadre of leaders who can consider the developmental level of the organization as they work actively to shift the culture may be necessary [15]. Different dimensions of leadership are required of organizational leaders. Acquiring skills and adopting perspectives markedly different from the skills and perspective required in the past will be needed in the change process [17].

Organizational culture is enduring. Creating stability and maintaining the *status quo* are key features of organizational culture. These are reasons why organizational culture is so difficult to change. In order for change to occur, the organization must first experience disequilibrium. A change in leadership, organizational restructuring, or organizational crisis are naturally occurring periods of disequilibrium in the culture. As such, these events in the life history of an organization offer an opportunity for change. At times, leadership may desire organizational change at a time when no naturally occurring event is expected. In this situation, leaders may need to precipitate disequilibrium by taking specific actions designed to "shake things up," creating a sense of urgency within members comprising the organization. Not recognizing the need for disequilibrium may be a key factor when an attempt to change the culture fails. In a review of ten case studies of academic health centers that were deliberate in attempting to reform the structure and management of the organization, several organizational leaders recognized that creating a sense of urgency among the faculty for change in the clinical enterprise was a critical element in the change initiative. However, these case studies focused on improving mission performance [18]. No systematic initiatives to improve or enhance faculty health through changing organizational culture are described.

Organizational disequilibrium can be created as impetus for change. Typically, this occurs in the form of planning for the future. A "strategic summit" or retreat, an administrative reorganization, or a job reclassification project may all be used to set the stage for a change initiative. During organizational disequilibrium, members of the organization become anxious. In an environment that may already be stressful, the resulting cognitive dissonance may "add fuel to the fire." Approaches that once brought success seem to result in failure. Too often, leaders and organization members regard these periods as wholly negative. Clearly, these are periods of organizational dissonance. However, disequilibrium is necessary even when the desired outcome is to create an improved organizational culture. Changing "the way we do things around here" offers the opportunity to resolve the organizational dissonance with a fresh approach that improves the overall quality of the organizational culture. Engaging the membership of the entire organization in changing the culture harnesses individual energy that may otherwise be expressed as anxiety and allows members to channel that energy toward positive change.

Promoting Faculty Health and Well-Being

A healthy workplace culture offers satisfying work relationships, opportunities for personal and professional growth, respect by leaders, control over workload, and expression of appreciation for successful performance. It is counterintuitive to think that organizations would *not* want to create a healthy workplace. Every academic health center has a need to retain faculty and reduce absenteeism. Recognition as a "Best Places to Work" organization is highly desired as this designation is a useful tool for recruiting faculty. Yet, the organizational culture of many of our academic health centers is not reflective of an environment where the health and well-being of the faculty and staff are a top priority. Are those of us in academic health centers incapable of "practicing what we preach to our patients?" A number of universities have implemented employee health and well-being programs offering services to faculty members.[3] While these programs are laudable,

---

[3] See http://www.med.umich.edu/wellbeing/; http://www.princeton.edu/pr/pwb/05/0131/6a.shtml; http://www.uhs.berkeley.edu/FacStaff/HealthMatters/ (Accessed April 3, 2008.)

they are oriented towards assisting individual faculty members and are not designed to change organizational culture [19]. While the needs of individuals should not be ignored, successful promotion of faculty health and well-being will likely require change in the organizational culture—change that will allow everyone in the academic health center to "feel genuinely valued and fulfilled" [20].

Leaders need to take several steps in the organizational change process in order to create a healthy workplace.

Make Use of Organizational Disequilibrium

Use the naturally occurring periods of disequilibrium created by changes in leadership, organizational restructuring, or organizational crisis, as opportunities to initiate change. If no naturally occurring event is expected, precipitate organizational disequilibrium by taking specific actions (i.e. strategic summit) to create a sense of urgency within members comprising the organization.

Assess the Culture

Assess the organizational culture. Although there are numerous measures available commercially, assessing the organizational culture may not be an easy task. The *Organizational Vital Signs* measure is one of many examples of commercially available tools that can be helpful in assessing culture [21]. On the other hand, an organization may wish to consider development of measures of faculty health and of a healthy workplace that consider the unique culture of the organization. Whether one chooses to create a new measure or to use an "off the shelf" measure, establishment of baseline measures followed by repeating the measures at timely intervals allows the organization to know whether the organizational change initiative is influencing faculty health.

The use of standardized or customized measures may not be sufficient. Retaining a consultant skilled in qualitative research methods, organizational psychology, or ethnography will supplement other measures. A skilled ethnographer will be able to "tease out" cultural nuances (through the use of a variety of techniques) that may have otherwise gone unnoticed. Standardized measures are often not sensitive enough to capture subtle aspects of the culture that may prove to be important as a strategy for change is developed. An approach that combines standardized measures with skilled observational measures is likely to yield more reliable results than the use of either measure alone. Similarly, McDaniel, Bogdewic, Holloway, and Jepworth suggest closely scrutinizing the academic health center and its policies through a "psychological health assessment" of the organization [22].

Examine Values, Vision, and Mission

Reexamine the organizational values, vision, and mission statements. If they are not consistent with the creation of a healthy workplace, they may need to be changed. Likewise, the way those values are applied across the organization may need to be examined. For example, many health care organizations, including academic health centers, include *compassion* as a core value. Does compassion apply to patients only? Or does it apply to the faculty members providing care? Certainly, faculty members working long hours and engaging in self-sacrifice serving patients are worthy of sympathetic concern for their suffering. In a discussion of the culture codes of doctors, nurses, and hospitals, Clotaire Rapaille points out the clash between the codes we apply to doctors—*hero*, to nurses—*mother*, and to hospitals—*processing plant*. It is critical to remember the negative consequences of organizational culture on those *heroes* and *mothers* working in *processing plants*.

Organizational values may need to reflect that the well-being of the healer, as well as the patient is a priority. Former AAMC President Jordan Cohen speaks of professionalism and humanism "not as separate attributes of a good doctor, but, rather, as intimately linked" [23]. While *professionalism* describes behavior, *humanism* implies a commitment to a set of deeply held beliefs about the intrinsic value of all human beings. Application of humanistic approach should not be limited to patient care. Indeed, self-care for the healer is a demonstration of the humanistic approach in action with regard to the healer, as well as to the patient.

Develop and Articulate a Compelling Vision of the Future

Articulate a compelling vision for the future of the organization as an organization offering satisfying work relationships, opportunities for personal and professional growth, respect by leaders, control over workload, and demonstrated appreciation for successful performance. A compelling vision for the future should include an environment where the humanistic approach is applied to the faculty, as well as to those served. Creating satisfying work relationships, opportunities for personal and professional growth, and respect by leaders are key features of such an environment.

Communicate

A key success factor in any attempt to change organizational culture is communication. Unless the members comprising the organization are committed to the change, it will not happen. Beginning with the articulation of a compelling vision of the future, leaders must be strategic in communicating with all of the members comprising the organization. Creating an "explicit communication strategy" is well worth the investment of time and resources [24]. This is a critical ingredient in the recipe for change as the next step in engaging the members of the organization. Members will not engage in the change process unless and until they are informed and know what is expected of them. Essentially, it is impossible to "overcommunicate" about what is expected of organization members. Consistent, simple, and repeated messages communicated through multiple channels over time are most likely to be understood. Hopefully, these messages will also inspire changes in behavior.

Engage Organization Members

Engage individuals within the organizational culture to collaborate in thoughts and action. Engaging individuals allows the organization to harness talent, perspectives, and energy and apply those to the change initiative. As leaders move the change process forward, both individual and organizational factors should be considered simultaneously if the desired outcome is to enhance health at the individual and organizational levels.

Conclusion

Organizational culture is enduring—but it is not immutable. Changing organizational culture leads to anxiety, and, very likely, to pain as members comprising the organization try to adapt to change. Mistakes will be made and, without doubt, things will be different as the organization copes with unforeseen situations [18].

A key ingredient in changing organizational culture is leadership. The nature and structure of academic health centers create many challenges for leaders and members comprising the organization. As organizations, academic health centers possess many of the characteristics of processing plants and do not lend themselves to promoting the health and well-being of the faculty. Creating organizational cultures within academic health centers that promote the health and well-being of faculty members requires deliberate action. Even with deliberate action, the process may be slow. As in any period of change, a high degree of ambiguity may persist within the organization and some members will have great difficulty tolerating the ambiguous circumstances. While individual temperament may predispose some faculty members to discomfort with organizational ambiguity, the ability to tolerate ambiguity is largely a learned behavior. Faculty members can—and will—tolerate changing the "way we do things around here" if leadership prepares and engages them in the change process. Ideally, the goal of leadership should be to create an organizational culture that not only promotes faculty health and well-being, but also inspires and nurtures *vitality* of the individual faculty members. In another chapter in this collection, *vitality* is defined as "a state in which there is optimal capability to realize shared individual and institutional goals" [25]. Inspiring and nurturing individual vitality creates organizational vitality—and *vice versa*. True organizational vitality derives from an organizational culture committed to the health and well-being of its faculty.

References

1. Schein, E.H. (1992) *Organizational Culture and Leadership* (2nd edition) San Francisco, CA: Jossey-Bass.
2. Bower, J.L. (1966) *The Will to Manage: Corporate Success Through Programmed Management* New York: McGraw-Hill.
3. Peterson, M., and Wilson, J.F. (2002) The culture-work-health model and work stress *American Journal of Health Behavior* **26(1)**, 16–24.
4. Grigsby, R.K. (2006) Managing organizational pain in academic health centers *Academic Physician and Scientist*, **January**, 2–3.
5. Dahlin, M.E., and Runeson, B. (2007) Burnout and psychiatric morbidity among medical students entering clinical training: a three year prospective questionnaire and interview-based study *BMC Medical Education* **7**(6) (Accessed October 23, 2007 at http://www.biomedcentral.com/1472–6920/7/6).
6. Maslach, C., and Leither, M.P. (1997) *The Truth About Burnout* San Francisco, CA: Jossey-Bass, 13–15.
7. Thorndyke, L.T., Gusic, M.E., George, J.H., Quillen, D.A., and Milner, R.J. (2006) Empowering junior faculty: Penn State's faculty development and mentoring program *Academic Medicine* **(81)**7, 668–673.
8. Brown, S., and Gunderman, R.B. (2006) Enhancing the professional fulfillment of physicians *Academic Medicine* **(81)**6, 577–582.
9. Thompson, N., Stradling, S., Murphy, M., and O'Neill, P. (1996) Stress and organizational culture *British Journal of Social Work* **26**, 647–665.
10. Peterson, M. (2004) What men and women value at work: implications for workplace health *Gender Medicine* **1(2)**, 106–124.
11. Aycan, Z., Kanungo, R.N., and Sinha, J.B.P. (1999) Organizational culture and human resource management practices: the model of culture fit. *Journal of Cross-Cultural Psychology* **30(1)**, 501–526.
12. Peterson, M., and Wilson, J.F. (2002) The culture-work-health model and work stress *American Journal of Health Behavior* **26(1)**, 16–24.
13. Williams, E.S., Konrad, T.R., Scheckler, W.E., et al. (2001) Understanding physicians' intentions to withdraw from practice: the role of job satisfaction, job stress, mental and physical health *Health Care Management Review* **26(1)**, 7–19.
14. Drath, W. (2001) *The Deep Blue Sea: Rethinking the Source of Leadership* San Francisco, CA: Jossey-Bass.
15. Kirch, D.G., Grigsby, R.K., Zolko, W.W., et al. (2005) Reinventing the Academic Health Center *Academic Medicine* **80(11)**, 980–989.
16. Rapaille, C. (2006) *The Culture Code* New York: Broadway Books.
17. Grigsby, R.K., Hefner, D.S., Souba, W.W., and Kirch, D.G. (2004) The future-oriented department chair *Academic Medicine* **79**, 571–577.
18. Griner, P.F., and Blumenthal, D. (1998) Reforming the structure and management of academic medical centers: case studies of ten institutions *Academic Medicine* **73**, 818–825.
19. Hubbard, G.T., and Atkins, S.S. (1995) The professor as person: the role of faculty well-being in faculty development *Innovative Higher Education* **20**, 117–128.
20. Kirch, D.G. (2007) Crossing the cultural divide in academic medicine *AAMC Reporter* **17(2)**, 2.
21. Institute for Organizational Performance *Organizational Vital Signs (OGS)*. (Accessed April 3, 2008 at http://www.eqperformance.com/ovs.php)

22. McDaniel, S.H, Bogdewic, S.P., Holloway, R., and Hepworth, J. (2007) The architecture of alignment: leadership and the psychological health of faculty *Proceedings of the faculty health conference* Houston, TX: University of Texas Health Sciences Center.
23. Cohen, J.J. (2007) Viewpoint: linking professionalism to humanism: what it means, why it matters *Academic Medicine* **82(11)**, 1029–1032.
24. Gilmore, T.N. (1999) *Sweeping People into a Campaign for Organizational Change* Philadelphia, PA: Center for Applied Research.
25. Viggiano, T.R., and Strobel, H.W. (2007) The career management life cycle: a model for supporting and sustaining individual and institutional vitality *Proceedings of the faculty health conference* Houston, TX: University of Texas Health Science Center.

Chapter 10
The Ethics of Self-Care

Craig Irvine

Abstract The medical academy's primary ethical imperative may be to care for others, but this imperative is meaningless if divorced from the imperative to care for oneself. How can we hope to care for others if we, ourselves, are crippled by ill health, burnout, or resentment? The self-care imperative, however, is almost entirely ignored by medical academicians and professional ethicists. Indeed, biomedical ethics focuses almost exclusively on the application of universal ethical principles to the treatment of patients and the protection of research subjects, discounting, even hindering, the introspection essential to the practice of ethical self-care. If we are to heed the self-care imperative, medical academicians must turn to an ethics that not only encourages, but even demands care of the self. We must turn to narrative ethics. Since narrative is central to the understanding, creation, and recreation of ourselves, we can truly *care* for ourselves only by attending to our self-creating stories. Narrative ethics brings these stories to our attention; so doing, it allows us to honor the self-care imperative.

Keywords Ethics clinical, ethics medical, ethics research, principle-based ethics, narrative medicine, narrative ethics

We are ethically obligated to care for ourselves. This, I believe, is incontrovertible. Our primary ethical imperative may be to care for others, but this imperative is meaningless, empty, if divorced from the imperative to care for oneself. Indeed, we may grant, with Emmanuel Levinas, that the imperative to care for others is *the* "primordial, irreducible, and ethical, anthropological category" [1, p. 158]. We may even grant that this imperative is so powerful, so fundamental, that one's response, as Levinas scholar Richard Cohen writes, "goes all the way to giving the very self of the self" [2, p. 294]. Yet I cannot give myself if I have no self to give. I must care

C. Irvine
Assistant Professor of Clinical Behavioral Science in the Center for Family and Community Medicine, College of Physicians and Surgeons, Columbia University, New York, USA
e-mail: ci44@columbia.edu

T.R. Cole et al. (eds.) *Faculty Health in Academic Medicine*,
© Humana Press, a part of Springer Science + Business Media, LLC 2009

for my hands, if I am to lift the fallen; my heart, if I am to love the stranger; my mind, if I am to cure the ill; my eyes, if I am to find the lost, and my soul, if I am to guide them home. No matter how it is conceived—philosophically, theologically, psychologically—the imperative to care for others is always already an imperative to care for myself.

Unfortunately, biomedical ethics focuses almost exclusively on the application of universal ethical principles to the treatment of patients and the protection of research subjects, thus discounting, even hindering, the introspection essential to the practice of ethical self-care. As I will show below, if we are to heed the self-care imperative, medical academicians must turn to an ethics that not only encourages, but even demands care of the self. We must turn to narrative ethics. Since narrative is central to the understanding, creation, and recreation of ourselves, we can truly *care* for ourselves only by attending to our self-creating stories. Narrative ethics brings these stories to our attention; so doing, it allows us to honor the self-care imperative.

Spaceship Ethics

For clinicians and researchers in academic medicine, professionally dedicated to addressing "the inevitable and preemptory ethical problem" of suffering (the other's demand for aid) [1, p. 158], the self-care imperative should be particularly compelling. Too often, however, this imperative remains entirely unheeded. As the demands on medical academicians grow greater and greater, evidence of burnout, ill health, and resentment also increases [3]. Much is written about the professional ethics of clinical care and medical research, but virtually none of this writing concerns the ethics of self-care. Rather, ethical discourse focuses on the treatment of patients and the protection of research subjects. Certainly, too much can never be written about the ethics of clinical care and the ethics of research, but unless attention is also paid to the ethics of self-care, ethical discourse for medical academicians is missing its essential third leg, without which it must remain perpetually off-balance and thus prone to collapse.

This lack of balance is perhaps most evident in the way the medical academy trains its initiates. Physicians- and researchers-in-training are generally offered little if any instruction in the care of the self, let alone its ethical implications. In fact, what they are taught, both implicitly, by observing their instructors' behavior, and explicitly, in courses on clinical conduct, is that they are to take themselves entirely out of the equation. Their personal thoughts, feelings, embodiment simply do not matter, do not figure at all in the clinical or research equation. Yet without honest self-reflection—without turning one's attention explicitly to one's body, emotions, thought—ethical self-care is impossible. Rather than learning to care for themselves, in ethically sound ways, students learn, in effect, that self-care is immoral.

This suppression of self-care was brought home to me in a particularly poignant way by a story written by a fourth-year medical student, Ashley, for an elective ethics course she took with me. Ashley's story was about an experience she had had almost two years earlier, as a third-year medical student, on the first morning of her first inpatient rotation. Early that morning, a patient named Mary was admitted to Ashley's hospital floor. Mary, who was not much older than Ashley, had been hospitalized with sepsis, caused by immune suppression from chemotherapy. Shortly after arriving on the floor, Mary developed Acute Respiratory Distress Syndrome. The entire team ran to her room, and the Chief Resident told Ashley to sit by the bed and encourage Mary to relax. For more than 5 hours, while residents and attendings ran in and out of the room doing everything in their power to arrest Mary's respiratory decline, Ashley held Mary's hand, repeating, over and over again, "Just breath. Relax, it's going to be okay. Breath. Please try to relax. We're all here for you. Just breath." When Mary stopped breathing, the Chief Resident pushed Ashley away from the bed, and he and the rest of the team began the code. Death was declared several minutes later. The team abruptly left the room, leaving Ashley alone with Mary's battered body. No one ever spoke to her about Mary's death.

When Ashley finished reading her story to me, she looked up and said, through her tears and without irony, "I just wish I'd been able to *do* something for Mary, like everyone else. I felt so helpless. Just useless and in the way." In the two years since Mary's death, Ashley had never shared this story with anyone at her school.

Ashley was well-schooled in the four "principles" of biomedical ethics: she understood her obligation to respect Mary's freedom of choice (autonomy), her obligation not to harm Mary in any avoidable way (nonmaleficence), her obligation to help Mary in every possible way (beneficence), and her obligation to provide all the care to which Mary was entitled (justice). Yet understanding all these principles did not help Ashley, in practice. Her ethics education was too abstract to help her work through this *particular* case—the specific, *lived* experience of her sense of responsibility to her patient and to herself. She understood how the principles of biomedical ethics apply to patients in general, but Mary was not a patient in general. Ashley's training offered no guidance or solace in working through her deep sense of moral failure, let alone her grief, following Mary's death.

The problem is not that Ashley's understanding of the four principles was inadequate; the problem is with the principles themselves. Or rather, with ethics curricula that rely exclusively on principle-based reasoning. The inadequacy of such curricula is increasingly manifest. As Rita Charon writes, "Over the past decade, conventional bioethics has struggled to find its way among its chosen principles and has found itself too thin to address adequately the actual value conflicts that arise in illness" [4, p. 208].[1] Barbara Nicholas and Grant Gillett argue that a principle-based ethics, or "principlism," not only raises "philosophical difficulties with criteria for application of the principles" and "problems with how one resolves conflict between principles," but also engenders "an unease among practitioners arising from the realization that the realities and practicalities of clinical practice are not paid sufficient attention" [6]. Indeed, as John O'Toole contends, principle-based

---

[1] See also [5].

ethics "lays siege to our very ability to act out the ethical dramas we will face daily as physicians," and thus "serves only to further disconnect the medical student from the human experience" [7]. This disconnection from human experience is twofold: it alienates medical students (and researchers, and residents, and all other medical academicians) both from the patient's experience and from their own.[2]

Biomedical ethics training thus fosters the medical academy's hostility to introspection, making it all the more difficult to conceptualize, let alone practice, ethical self-care.[3] By employing abstract, one-size-fits-all, deductive reasoning, which argues from general principles down to particular cases, principlism teaches medical academicians to master ethics "cases" in the same way they master proofs in logic or "problems" in biochemistry, physics, and statistics. Principlism reinforces a sense of invulnerability, of detachment, of domination, imposing a "unifying general view," as Arthur Frank puts it [9, p. 13], on all experience and thus bringing it under control. In this regard, Frank finds William James's self-condemnation of his own "unholy" ambition particularly revelatory:

> I am convinced that the desire to formulate truths is a virulent disease. It has contracted an alliance lately in me with a feverish personal ambition, which I never had before, and which I recognize as an unholy thing in such a connection. I actually dread to die until I have settled the Universe's hash in one more book ...! Childish idiot—as if formulas about the Universe could ruffle its majesty. [10, p. 1344]

James's "disease," Frank contends, is characteristic of the "modernist universalism" of the medical academy: "To settle the Universe's hash is to place oneself outside the vulnerability and contingency that being in the Universe involves. The intellectual infected with such an ambition ceases to think of himself as a body, thus disclaiming the vulnerability that bodies share" [9, p. 19]. Unfortunately, disclaiming one's vulnerability does not make one invulnerable. Indeed, ceasing to think of oneself as a body only makes it that much more difficult to address the injuries, illnesses, stresses, weaknesses, and exhaustion that we all, as embodied beings, must inevitably face. The modernist universalism of biomedical ethics thus magnifies the harms we suffer as embodied beings: refusing to acknowledge them, we only make them worse, allowing them to fester and grow, unchecked. Modernist universalism, Frank writes, is itself a diseased drive—the drive to be "the beginning and end of all things": this is the "weight modernity puts on its heroes. Physicians feel this weight" [9, p. 153]. Far from questioning the morality of modernist heroics, and thus helping to lift its burden, ethics, as traditionally practiced in the medical academy, piles on more weight. By adopting the universalist view of modernism, prinicplist ethics places us above, outside of, beyond the specific, lived reality of our embodied experience. It turns us away from self-care, thus making us all the more unhealthy.

---

[2] As Suzanne Poirier writes, "The inability of many ethics discussions to address the personal values or feelings that often undergird ethical dilemmas suggests that the desire to deny or bury the personal voice under a formalized professional one may limit expression that is valuable to the conversation of bioethics" [8, pp. 57–58].

[3] "Professional culture has little space for personal becoming. Young doctors are not trained to think of the careers ahead of them as trajectories of their own moral development" [9, p. 159].

Asked to assume a burden impossible to bear, physicians and researchers develop coping mechanisms, Frank asserts, that "can warp an otherwise decent mind" [9, p. 147]. He demonstrates this with a quotation reported by Charles Bosk. During his research on physicians who are genetic counselors, Bosk asked one of his subjects how he "came to grips with all the 'accidents' or 'mistakes' [of medical practice] that he saw." Following is the physician's response:

> What you have to do is this, Bosk. When you get up in the morning, pretend your car is a spaceship. Tell yourself you are going to visit another planet. You say, "On that planet terrible things happen, but they don't happen on my planet. They only happen on that planet I take my spaceship to each morning". [11, p. 171]

The implication, of course, is that the "terrible things" happening on the hospital planet trouble only its natives (the patients), not the physicians, who are, after all, only visitors. Physicians believe that they are protected from all that happens there, shielded by the principles of biomedical ethics, which we might now rename "spaceship ethics."[4] These principles allow physicians like the one interviewed by Bosk to place themselves outside of their patients lives. Indeed, the spaceship they travel in is the universalist perspective we examined above. The irony is that these universal principles of biomedical ethics, as we saw, only serve to increase the burden of modernist heroics, an increase that drives physicians to take further refuge in the principles, still further increasing the burden. It is a vicious circle—a relentless, ever widening orbit between two hostile planets (self/home vs. other/work). Once they climb into their spaceships, medical academicians are doomed to a life unmoored— lost in space—as they drift further and further away both from themselves and from others. It is no wonder Ashley was left to her own devices: the rest of the crew was already in orbit. Their behavior was alien to Ashley (and Ashley's needs alien to them) because she had not yet learned to leave her *self* behind; consequently, she could not yet pretend that she had not been affected by Mary's death.

Narrative Ethics

Given the predominance of spaceship ethics, how could Ashley hope to learn to care for herself? How could she face the traumas of her work—face them directly, honestly, *ethically*—if her work is utterly separate from her life, alienated from the familiar, *homely* life of her feelings, her embodiment?[5] Is she doomed to climb on

---

[4] "According to modernist universalism, the greatest responsibility to *all* patients is achieved when the professional places adherence to the profession before the particular demands of any individual patient. Such professionalism—paradigmatic of modernity—is responsible less to individual people than to truth " [9, pp. 15–16]. Of course, the "individual people" that professionalism holds hostage to the "truth" includes the professionals themselves.

[5] As Martha Nussbaum writes, "Our cognitive activity centrally involves emotional response. We discover what we think … partly by noticing how we feel; our investigation of our emotional geography is a major part of our search for self-knowledge" [12, p. 15].

board—doomed to a voyage utterly alien to self-care—or is there an alternative to this cold, orbital drift into space?

The answer, I believe, lies in Ashley's story—or rather, in her act of storytelling itself. In this act we discover an ethical alternative that not only encourages, but even demands care of the self. In storytelling, we discover narrative ethics, an ethics that "lightens the load on people," [9, p. 153] that is "reciprocal and reflective, [demanding] vision and courage, all the while replenishing one's store of vision and courage" [13, p. xii].

Stories are the primordial means through which we make sense and convey the meaning of our lives. Our very selves, Frank writes, "are perpetually recreated in stories. Stories do not simply describe the self; they are the self's medium of being" [9, p. 53]. As the universalist notions of modernism are challenged by postmodern thinkers, the power of narrative to create, express, and recreate the *unique* reality of the self—its irreducible, inassimilable particularity—has become increasingly manifest: *"Postmodern times are when the capacity for telling one's own story is reclaimed* ... when people's own stories are no longer told as secondary but have their own primary importance" [9, p. 7]. This primacy is acknowledged across diverse disciplines. Indeed, "[h]istorians, cognitive psychologists, social scientists, theologians, psychiatrists, and literary critics," Charon and Montello note, have all "come to recognize the central role that narrative plays in the way we construct knowledge, interpret experience, and define the right and the good" [13, p. x].

If narrative is central to our construction of knowledge, our interpretation of experience, and our definition of the right and the good, it stands to reason that it is indispensable to moral inquiry [9, 13]. One cannot hope, after all, to recognize, illuminate, and resolve moral dilemmas without considering the narratives—the very medium of our *being*—in which those dilemmas *live* [14]. A biomedical ethics based in narrative would engage every level of our being, including and perhaps especially our emotional life, immeasurably deepening our moral sensitivity.[6] Narrative bioethics, therefore, "would look beyond a calculus of principle and reason. It would require us to account for the emotion so crucial to ethical action and to the ways in which stories work on us" [15, p. 210].[7]

As we saw above, Ashley could not understand the meaning of her experience at Mary's bedside until she had written and shared her story. Before this, Ashley's moral view was circumscribed by the universalist principles of her medical training—principles with which she could not possibly make sense of her role in Mary's

---

[6]"Narrative is often indispensable in helping us grasp what our deepest values are. ... In contrast to principle alone, narrative in its detailed, emotion-rich representation of experience can help us recognize implicit values and negotiate conflicts of moral action" [15, pp. 212–213] See also [16].

[7]Narratives, Hilde Lindemann Nelson writes, "cultivate our moral emotions and refine our moral perception; ... they teach us our responsibilities; they motivate, guide, and justify our actions" [17, p. 47].

death. Writing and reading her story, as I'll make clear below, allowed Ashley to re-interpret her role, to "define the right and the good," to *value* her experience in the context of a much broader ethical horizon—a horizon that embraced her particular, lived, *emotional* experience as a medical student, a young woman, a caring human being.

Looking beyond principle and reason, we find the wellspring of meaning, in all its glorious complexity, ambiguity, superfluity. This narrative source is beyond the grasp of our technical mastery. We can never get behind or above the act of story-telling, can never obtain a comprehensive view, can never fully account for it, master it, because it is the very condition of all accounting, all meaning. Attending to narrative, we challenge the technical mastery through which we dominate reality, and we engage questions of right and wrong with humility, nurturing our *connection* both to others and to ourselves—to our shared humanity. Sayantani DasGupta has coined the phrase "narrative humility" to highlight the way narrative moves us out of the mastery so damaging to ourselves, our patients, and our research subjects: "Narrative humility," she writes, "acknowledges that our patients' stories are not objects that we can comprehend or master, but rather dynamic entities that we can approach and engage with, while simultaneously remaining open to their ambiguity and contradiction, and engaging in constant self-evaluation and self-critique about issues such as our own role in the story, our expectations of the story, our responsibilities to the story, and our identifications with the story—how the story attracts or repels us because it reminds us of any number of personal stories" [18, p. 981]. Narrative ethics, unlike spaceship ethics, is therefore inseparable from everyday life [13]. This does not mean, as Jerome Bruner notes, that the proponents of narrative ethics hope to "detechnicalize sickness or health care." Obviously, no one aims to hinder the scientific progress of medicine's highly technical understanding and treatment of disease. But for health care to progress *ethically*, attentive to the care of others and the self, "you've got to rehumanize it as well—relate it to life. Who on earth," asks Bruner, "wants to practice like a robot? Or turn their patients into robots?" [19, p. 8].

Sadly, practicing "like a robot" often entails treating stories robotically. Typically, medical professionals "understand stories as something to carry a message away from—as in, 'What did you learn from that history?' The professional, as paradigmatic modernist, is always moving on, the sooner to get to the next thing and move on from that" [9, p. 159]. In contrast to this mechanistic approach, which treats stories as objects from which information is extracted by "superior," nonnarrative, cognitive tools, narrative ethicists recommend that we learn to think *with* stories:

> Not think about stories, which would be the usual phrase, but think with them. To think about a story is to reduce it to content and then analyze that content. Thinking with stories takes the story as already complete; there is no going beyond it. To think with a story is to experience it affecting one's own life and to find in that effect a certain truth of one's life. [9, p. 23]

As Frank makes clear, the way that stories create and communicate meaning is primary and complete. There is not a more "fundamental" way of thinking that must be applied to stories in order to reveal their ethical import.[8] On the contrary, storytelling itself is ethic's invocation and attentive listening ethic's response. David Morris puts it thusly: "Thinking with stories is a process in which we as thinkers do not so much work on narrative as take the radical step back, almost a return to childhood experience, of allowing narrative to work on us" [15, p. 196]. It is impossible to categorize or to comprehend all of the ways stories work on us; one cannot reduce their content to analyzable data, subsumed under universal principles. As soon as we apply principles to stories, we lose their primary meaning, a primacy that includes the invocation of emotion as integral to ethical self-understanding [21]. "The first lesson of thinking with stories," Frank contends, "is not to move on once the story has been heard, but to continue to live in the story, becoming in it, reflecting on who one is becoming, and gradually modifying the story. The problem is truly to listen to one's own story, just as the problem is truly to listen to others' stories" [9, p. 159].

This final point of Frank's—that to practice ethics one must truly listen to one's own story—is key to understanding narrative's role in the ethics of self-care. Since narrative is central to the understanding, creation, and recreation of ourselves, we can truly *care* for ourselves only by attending to our stories. Our needs, desires, and aspirations are given both form and content—integrated and reintegrated into the continual unfolding of ourselves—through the self-creating power of narrative. Narrative, Joanne Trautmann Banks writes, "inevitably expresses and transforms who we are at every level of our being" [22, p. 219].[9] Through my self-story, I decide, express, and enact who I will be.[10] I therefore take responsibility for myself through active, self-conscious engagement in the narration of my story.[11] What Arthur Frank writes of ill people applies equally to physicians and researchers: "Postmodern illness stories are told so that people can place themselves outside the 'unifying general view.' For people to move their stories outside the professional purview involves a profound assumption of personal responsibility" [9, p. 13]. In postmodern times, Frank contends, "becoming a narrator of one's own life implies an assumption of responsibility for more than the events of that life. Events are contingent, but a story can be told that binds contingent events together into a life that has a moral necessity" [9, p. 176].

---

[8] "I am interested," Howard Brody writes, "in what Hilde Nelson has classified as telling, comparing, and invoking stories; that is, using narrative as part of moral reasoning, rather than merely as illustrative of moral conclusions derived from other methods of reasoning." [20, p. 149]

[9] As Hilde Lindemann Nelson puts it, narratives "make intelligible what we do and who we are; ... through them, we redefine ourselves" [17, p. 47].

[10] "The self-story is not told for the sake of description, though description may be its ostensible content. The self is being formed in what is told" [9, p. 55].

[11] "What the story teaches is that there is always another story, and other stories have always been possible. One meaning of this lesson is that life is lived in decisions, each setting in place a different way of telling the story. Because these decisions have consequences—the plot cannot be reversed at will at any point—they are moral. Thinking with stories means that narrative ethics cannot offer people clear guidelines or principles for making decisions. Instead, what is offered is permission to allow the story to lead in certain directions" [9, p. 160].

Storytelling's ability to grant an individual life moral coherence is restorative, healing.[12] This healing, however, is not just personal, but interpersonal as well: "Because stories can heal, the wounded healer and wounded storyteller are not separate, but are different aspects of the same figure" [9, p. xxi]. Narrative ethics recognizes that all of us, at one time or another, will become fractured—whether by illness or by professional setbacks, stress, disappointments, or even success: "Sooner or later," writes Frank, "everyone is a wounded storyteller" [9, p. xiii]. Narrative ethic's recognition of this shared vulnerability undermines modernist heroics, with its spaceship ethics that so effectively, and damagingly, alienates healers from ill people, from other healers, and from themselves.

Writing, reading, and discussing the story of one's alienation, therefore, are often the first steps in overcoming it. This was certainly true for Ashley. During our discussion, we considered the role the "character" of the medical student plays in the story of Mary's death. In this story, Ashley discovered, the student plays a much more important role than any of the doctors: Mary would have died whether or not Ashley was there, but her death would have been far less peaceful. While the importance of Ashley's role seemed immediately obvious to me, as it would to most readers of her story, Ashley had not, previously, been encouraged to acknowledge the moral authority of her actions. On the contrary, her professional training had actively discouraged this acknowledgement. Every death, for medicine, is simply a defeat—end of story—at the hands of its worst enemy. Writing and sharing her story offered Ashley a means of memorializing Mary's death—a means of preserving the memory, and so *placing* the meaning, of an event radically dislocated by her professional training. This was healing for Ashley. Reflecting on the story of her life, of Mary's life, of *their lives*, Ashley heeded the imperative to care for herself.

Obviously, writing and relating the story of Mary's death could not possibly repair all of the psychic, and ethical, injuries Ashley suffered during her medical training. It was, however, an essential first step in what must be a continuous, ever-evolving narrative process: for Ashley, as for all health professionals, healing requires an ongoing commitment to narrative self-reflection. This commitment goes beyond the reparation of injuries wrought by the modernist heroics of the medical academy. Indeed, narrative ethics requires and promotes a fundamental change in the very culture of the medical academy itself.

This cultural change, I believe, can be effected in every academic setting in which medicine is researched and practiced. I am confident that this is true, because I have witnessed narrative's efficacy in what is perhaps the environment most hostile to any such transformation: the inpatient wards. It is on the wards that medical professionals are the most rushed, the most anxious, the most focused, and so the most in thrall to medicine's modernist heroics. Yet in this setting I have held Narrative Medicine Rounds every other Friday morning for the past seven years. The Family Medicine inpatient team of the New York-Presbyterian hospital consists

---

[12] "According to Kleinman, narrative brings order and cohesion to suffering" [16, p. 131].

of two attendings, one first-year resident and one second-year resident, two interns, and one or two medical students. We begin rounds by together reading a short essay, story, or poem by a professional author. We then discuss the ways the author constructs and conveys meaning in her or his text. Following this discussion, I ask the team to write about their inpatient experiences, keeping in mind the narrative lessons we have learned in our discussion. When they have finished writing, the team reads their stories aloud, and we discuss them in much the same way as we discussed the opening texts.

Over time, the culture on the Family Medicine wards has shifted from one that, at best, tolerated thinking *about* stories to one that now fully welcomes thinking *with* stories. The benefits of this change are striking. On a recent Friday morning, for example, one of the attending physicians, Marion, asked if she could read a story to the team that she had written the previous evening. Marion has been with our program for many years, having completed her residency training at Columbia before joining our faculty. As an intern, Marion had been utterly focused on learning the medicine, the whole medicine, and nothing but the medicine. Like most interns, she was understandably resistant to anything that she felt might distract from this focus; all such distractions seemed to threaten her ability to keep her patients alive. Now, as a young attending, she is one of the faculty members most gifted at thinking with stories. I was therefore happy to set aside Hemingway's "Hills Like White Elephants," which I'd planned to read with the team that morning, so that Marion could read the following story:

> *"My mother is dying" the voice says. I must go see her. My family calls her "the fruit lady." She always brings a heavy bag of fruit, pineapples peeled ever since I told her I was too lazy to peel them. I can see her red lipstick and thick, straight, salt and pepper hair. Before the chemo, that is. There was something young and alive about her, even sexy, despite her 60 years.*
>
> *"They put in a chest tube to drain the fluid and her lung collapsed. The other was full of cancer," her daughter explained. I had hoped this procedure would let her walk around without oxygen during these final few months.*
>
> *I had forgotten that she had smoked many years ago, when she first complained that her throat felt tight, her heart was racing. It took a few visits to realize this was not a virus or a sour stomach. I told her right away, feeling encased in plexiglass during that conversation, trying to convey facts.*
>
> *She rallied with her usual grace and we had two more years of warm, intimate visits, stories of her family and friends, bags of fruit. And now she was dying.*
>
> *On the train there were lots of people shoved into the car, going home after a long day of work. What will I say to a room full of her family, even to her? What explanation can I offer them about why medicine cannot stop this process, this pain? How will I say goodbye, in Spanish, no less?*
>
> *Presbyterian Hospital subway stop—so familiar, but somehow different tonight. On the way into the hospital, lots of important looking people—surely they have saved lives today. I am going to watch one end.*
>
> *How does one say "goodbye" in Spanish? I racked my brain. Hasta luego, nos veimos, si "dios quiere"… what else? I seriously wonder if this is a language that does not have a word for a final goodbye.*
>
> *The room is full of people. She is blue and in pain, struggling with each breath, her oxygen sat is in the 50 s. I hold her hand and swallow my heart as she says, "Gracias por todo, doctora, te quiero mucho". I whisper in her ear that I will think of her whenever I eat pineapple, that I love her too. I think she smiles.*

The oncologist is gone for the day. I page the intern to double the morphine drip and cancel the CT scan that surgery requested to check the placement of the chest tube.

The subway is emptier on the way home. I feel oddly peaceful. I open the door to my apartment and see simple signs of life. Everyone looks pink and healthy. I breathe a sigh of relief. My husband is puttering around the kitchen. "How do you say goodbye in Spanish?" I ask.

He looks at me incredulously and says "Adios." [23]

Marion's story, I believe, highlights beautifully the benefits of a sustained and sustaining culture of ethical self-care. To nurture this culture, one must offer forums for clinicians and researchers to write, share, and reflect on their own self-telling narratives, like the inpatient narrative medicine rounds described above. While the benefits of this cultural change are numerous, I would like to focus on the four that are most clearly in evidence in Marion's story and in the team's response the morning it was shared: (1) attention to singularity, (2) heightened awareness of narrative temporality, (3) strengthening of community, and (4) greater acceptance of death.

Singularity

Marion's story is not universal. Unlike the principles of biomedical ethics, this story about the relationship between Marion and the "fruit lady" does not *apply* to all relationships between all doctors and all patients. If any other doctor wrote a story about this patient, it would be a different story with an entirely different meaning. If Marion wrote a story about any other patient, it would be a different story with an entirely different meaning. She is not just any doctor, filling a role with any patient. As her story reveals, Marion is *this* doctor, caring for *this* patient.[13]

It is precisely narrative's focus on singularity that makes possible and structures all true moral deliberation,[14] including deliberation about care of the self. Once again, what Frank writes about ill people applies equally to physicians and researchers: "My concern is with ill people's self-stories as moral acts, and with care as the moral action of responding to those self-stories" [9, p. 157]. Moral activity has two sides: care given to the other in response to the other's self-story and care

---

[13] "Narrative knowledge addresses the particulars of the case, attends to multiple perspectives, and depends on descriptions of a specific or situated context and personal details that define meaning. It allows many possibilities to be present at once and recognizes the emotional aspects of living one's life" [24, p. 143].

[14] "Narrative approaches to ethics recognize that the singular case emerges only in the act of narrating it and that duties are incurred in the act of hearing it" [13, p. ix].

"By the ethical dimension," Kenneth Burke writes, "I have in mind the ways in which, through language, we express our characters, whether or not we intend to do so. … [W]e could say that language reflects the 'personal equations' by which each person is different from any one else, a unique combination of experiences and judgments. Thus there is a sense in which … each poet speaks his own dialect" [25, p. 28].

given to myself in response to my own self-story. By telling and attending to her unique story, Marion is called to respond to the self-care imperative.[15] She is called to examine who she is becoming *as a person, as a unique individual*—not as a substitutable role. This examination necessarily entails asking whether her actions are good or bad for herself. As Barry Hoffmaster contends, the "crucial test of a story might be the sort of person it shapes" [26, p. 1161]. This test helps Marion to understand how she might care for herself and thus become the person she wants to be. Michael White writes that a person often finds herself "situated in stories that … she finds unhelpful, unsatisfying, and dead-ended, and … these stories do not sufficiently encapsulate the person's lived experience or are very significantly contradicted by important aspects of the person's lived experience" [27, p. 14]. To care for oneself, therefore, one must take responsibility for one's narrative situation: "[P]ersons give meaning to their lives and relationships," White argues, "by storying their experience. … [I]n interacting with others in performance of these stories, they are active in the shaping of their lives and relationships" [27, p. 13].[16] Marion thus cares for herself by bringing her story into alignment with her own, unique experience.[17] As Marion's story demonstrates, it is possible to achieve this alignment even in the most "modernized" of medical environments.

Temporality

Marion's story, like all narratives, is a temporal configuration of events. This configuration is not strictly chronological. On the contrary, Marion begins in the story's "present" (*" 'My mother is dying' the voice says"*); flashes back to the past (*"Before the chemo, that is"*); returns to the present (*"They put in a chest tube to drain the fluid and her lung collapsed"*); moves back to a moment even further in the past (*"I had forgotten that she had smoked many years ago"*); narrates forward from this past to a point in the present just after the phone call (*"On the train there were lots of people shoved into the car"*); then narrates from this point forward through the hospital visit (*"The room is full of people"*), ending with the scene at home (*"I open the door to my apartment and see simple signs of life"*). This temporal configuration is crucial to Marion's ability to

---

[15] "The moral imperative of narrative ethics is perpetual self-reflection on the sort of person that one's story is shaping one into, entailing the requirement to change that self-story if the wrong self is being shaped" [9, p. 158].

[16] "Self-stories are told to make sense of a life that has reached some moral juncture" [9, p. 161].

[17] "Narrative ethics is compete, within its sphere. This sphere is not clinical adjudication but personal becoming. Narrative ethics is an ethics of commitment to shaping oneself as a human being. Specific stories are the media of this shaping, and the shaping itself is the story of a life" [9, p. 158].

make sense both of her care for her patient and her care for herself.[18] By bringing together events that are not contiguous in time, Marion's story sparks a flash of illumination that brings to light the meaning of her experience.[19] As Rita Charon puts it, "We learn who we are backwards and forwards, early memories taking on sense only in the light of far later occurrences and contemporary situations interpretable only in the web of time" [30, p. 67]. This temporal interpretation, Hilde Lindemann Nelson argues, is much more conducive to ethical reflection than the atemporal standpoint of principlism: "[U]nderstanding how we got 'here' is crucial to the determination of where we might be able to go from here, and this is where narrative is indispensable. ... Because narrative approaches ... move backward and forward, they are better suited to ethical reflection" [17, pp. 39–40]. Ethical self-reflection demands that we understand events *in time*. Principlist ethics seems to consider events as if they existed *outside* time—as if moral dilemmas could be understood without considering the past out of which they arise, the present in which they appear, and the future toward which they move. As Marion's story demonstrates, it is impossible to honor the imperative to self-care unless one attends to the narrative temporality of one's own self-telling stories. "Constructing time-bound, causal patterns," Charles M. Anderson and Martha Montello write, "enables us to make sense of primary experience. The narratives we build shape the ways we come to know ourselves and each other and create the symbolic space within which we make all our moral choices" [31, p. 85].

At the end of her story, Marion returns to her home. There, she feels "peaceful," she observes all the "signs of life," and she "breathes a sigh of relief." Surrounded by her loving and beloved family, she literally learns how to say goodbye to her patient. Like all good endings, this return to her home brings closure to Marion's story—a closure that gives to her experience, and so to her very self, a healing *wholeness*.[20] The ending of her story shapes the meaning of the many years of Marion's experience caring for her patient. It brings that experience to a close in a way that allows her to integrate it into her ever-unfolding life story—to make it a part of the whole of her life, rather than something separate, outside of her life. Paradoxically, telling a story (her configuration *of* time) about her search for coherence (her experience *in* time) enacts the coherence

---

[18] "Life moves on, stories change with that movement, and experience changes. Stories are true to the flux of experience, and the story affects the direction of that flux" [9, p. 22]. "Plot links the complicated and uncertain conditions that may—or may not—be essential ingredients in the events being represented, and then plot traces those conditions over time" [28, p. 77].

[19] "Memory is not only restored in the illness story; more significantly, memory is created" [9 p. 61]. David Carr argues that "at no level, and certainly not at the scale of the life-story itself, is the narrative coherence of events and actions simply a 'given' for us. Rather it is a constant task, sometimes a struggle, and when it succeeds it is an achievement" [29, p. 96].

[20] As Tod Chambers and Kathryn Montgomery write, "If plots are driven by their endings, then what is considered a proper ending will have a profound impact on the story that is told. The sense of an ending that can bring closure to moral problems is arbitrary, perhaps temporary, and always a human construction. These limitations are not negative features but rather, when fully recognized, the very business of bioethics" [27, p. 84].

for which she is searching.[21] This temporal paradox is productive, for it is the paradox through which a narrative becomes the story of why we tell ourselves stories.[22] In Marion's case, the fruit of her story, if you will, *is* the story of her fruit.

Community

"Only connect," E. M. Forster writes in the epigraph to his 1910 novel Howards End [32]. This admonition, I believe, is a vital expression of the self-care impera-tive. As Bruner puts it, "Everywhere you look, you run into the recognition of the fact that a human plight is never an island unto itself. So, what should you do? You connect" [19, p. 9]. Indeed, Samuel Shem writes, "the primary motivation of human beings is the desire for connection. … [T]he seeds of human misery are planted in disconnections, violations, isolation, and domination, and the core of healthy growth is the movement from isolation toward connection" [33, pp. 43–44]. Since the work of establishing interpersonal connections is fundamental to addressing "human misery," ethics is nothing if not this interpersonal work: "Perhaps the great-est life-value of ethics," writes György Lukács, "is precisely that it is a sphere where a certain kind of communion can exist, a sphere where the eternal loneliness stops. The ethical man is no longer the beginning and the end of all things, his moods are no longer the measure of the significance of everything that happens in the world. Ethics forces a sense of community upon all men" [34, p. 57].

Narrative honors this centrality of community to ethics—particularly to the eth-ics of self-care. In this regard, it is the antithesis of spaceship ethics, which actually makes a virtue of remaining *un*-connected, of fostering isolation.[23] Marion's story beautifully illustrates the power of narrative to transcend this isolation. Her story expresses and enacts a marriage between her work and home lives. Like all mar-riages, it is defined—as only narrative can reveal—by the web of relationships that comprise and support it. Marion's story reveals her active, self-reflective engagement

---

[21] As Frank writes, "The truth of stories is not only what was experienced, but equally what becomes experience in the telling and its reception. The stories we tell about our lives are not necessarily those lives as they were lived, but these stories become our experience of those lives" [9, p. 22].

[22] For Frank, "The Good story ends in wonder, and the capacity for wonder is reclaimed from the bureaucratic rationalizations of institutions, medicine. Being available to yourself ultimately means having the ability to wonder at all the self can be" [9, p. 68].

[23] "[T]he moral life cannot be conceived apart from one's relationships with others—a claim that contrasts to modern notions of the self as an isolated unit and that challenges the notion of the moral life as guided by abstract ideas, rules, and principles" [35, p. 73].

In contrast to pinciplist ethics, "the narrative approach … sees morality as a continual interper-sonal task of becoming and remaining mutually intelligible. In this view, morality is something we all do together, in actual moral communities whose members express themselves and influence others by appealing to mutually recognized values and use those same values to refine understand-ing, extend consensus, and eliminate conflict" [17, p. 46].

in making interpersonal connections and fostering narrative coherence. She shows how this coherence undergirds her familial and professional lives—how she returns to her family to find the words to say goodbye to her patient. Yet the story itself fosters the very connections, the very coherence, which is its principal theme. Marion's is therefore both a story *about* self-care and an act *of* self-care.[24] Yet again, Frank's reflections on illness apply to the self-care of both patients and medical academicians: "Serious illness is a loss of the 'destination and map' that had previously guided the ill person's life: ill people have to learn 'to think differently.' They learn by hearing themselves tell their stories, absorbing others' reactions, and experiencing their stories being shared" [9, p. 1].

Reading her story to *this* audience,[25] Marion fosters both the interpersonal connections narrated *in* her story (her family, her patient, and her patient's family) and the interpersonal connections created *by* her story (the inpatient team). As Julia E. Connelly writes, "Narrative knowledge allows and encourages human connections. One shared story triggers the telling of other stories by involved listeners, facilitates memories and personal reflections on past experiences, if only silently revealed, and creates an expanded awareness of the moment, including a recognition of the power of personal presence and connectedness" [24, p. 145]. This "expanded awareness" was certainly in evidence the morning that Marion read her story. When Marion finished reading, we discussed the story's meaning and impact. Every member of the team—from third-year medical student through well-seasoned attending—was inspired to share a story from her or his own experience.[26] In each of these stories the team discovered and created connections to all of the others, thus building upon and reinforcing the shared narrative coherence that characterizes a true community of storytellers. As Frank writes, "Listening is hard, but it is also a fundamental moral act; to realize the best potential in postmodern time requires an ethics of listening. … [I]n listening for the other, we listen for ourselves. The moment of witness in the story crystallizes a mutuality of need, when each is *for* the other" [9, p. 25]. The recognition of this mutuality of need shatters the illusion of modernist heroics—the "illusion of oneself as the beginning and end of all things." Once this illusion of self-sufficiency is shattered, "openness to communion is all that is left" [9, p. 154]. In this openness, we heal the wounds of our heroic isolation.[27]

---

[24] "If we assume the truth of the relational self, ethicists receive a personal dividend as they attend to one another's interpretations of the patient's story—their work contributes fundamentally to their own self-development" [22, p. 224].

[25] "It is one contribution of narrative theory to bioethics to recognize that moral discernment is a form of narrative activity and that arguments and principles are made within the specificity of particular narratives told by particular tellers to particular listeners" [35, p. 26].

[26] "People tell stories not just to work out their own changing identities, but also to guide others who will follow them. They seek not to provide a map that can guide others—each must create his own—but rather to witness the experience of reconstructing one's own map. Witnessing is one duty to the commonsensical and to others" [9, p. 17].

[27] "In the reciprocity that is storytelling, the teller offers herself as guide to the other's self-formation. The other's receipt of that guidance not only recognizes but *values* the teller. … Telling stories in postmodern times, and perhaps in all times, attempts to change one's own life by affecting the lives of others" [9, pp. 17–18].

Death

There is no greater challenge to modernist heroics than death. In death we face the absolute limit of our power, that over which, ultimately, we have no control. At best, we may be able to forestall death, but we can never prevent the inevitable: death *will* have its day. Yet medicine frames death as its enemy then refuses to acknowledge what this framing implies: according to its own terms, medicine's rate of failure is 100%. As Zoloth and Charon write, "Clinical and research medicine long ago erected death as the enemy, death as defeat. Through terror, perhaps, or through hubris, medicine endorsed the haunting illusion that it could conquer death. Despite the efforts of palliative care and hospice, attempts to understand and sometimes welcome death are marginalized and exert little influence on medicine's mainstream work—the big business of intensive care, clinical trials, and pharmaceutical and surgical intervention—all of which are predicated on finding the right triumphal ending for the narrative of illness" [36, p. 29]. Maintaining this illusion of its ultimate triumph, medicine denies itself the possibility of finding meaning in death.[28] It therefore also denies itself the possibility of finding meaning in life: how can one find meaning in life, after all, if one refuses to face the single most important fact of our temporal—temporary—existence. As we saw in Ashley's story, the effects of this denial are devastating: to deny death is to deny our very selves, thus making self-care impossible.

It is the task of ethics to move medicine out of this denial.[29] Indeed, ethics "is the profession committed to facing death directly," Zoloth and Charon argue, "calling out its name, and finally, unlike medicine, finding meaning in the lost places in the clinical world where death has established its dominion" [36, p. 29]. As we saw above, principlist ethics often reinforces medicine's modernist heroics, colluding in its denial of death. For narrative ethics, on the other hand, "death is the issue, the topic, the thing itself" [9, p. 29]. Narrative ethics is concerned with the whole story.[30] It does not approach the meaning of a story from a perspective outside or above it. Rather, it finds meaning *inside* the story *as a whole*. The death of a patient, research subject, loved one, therefore, does not lie outside the story. In Marion's case, for example, death is integral to, even revelatory of, the meaning of her narrative. Creating a memorial in honor of death—integrating death into the whole of

---

[28] "Modernity, Bauman argues, exorcises the fear of mortality by breaking down threats, among which illness is paradigmatic, into smaller and smaller units. To use May's distinction, the big mystery becomes a series of little puzzles. Medicine, with its division into specialties and subspecialties, is designed to effect this deconstruction" [9, pp. 83–84].

[29] "Modernity disallows any language other than survival; the modernist hero cannot imagine any other way to be, which is why physicians are often genuinely baffled by criticisms. People in postmodern times need different languages of meta-survival with various messages that death is all right. Clinical ethics needs these messages" [9, p. 166].

[30] "Modernist medicine has regarded suffering as a puzzle to be 'controlled' if not eradicated. Postmodern illness culture, lay and medical, recognizes a need to accept suffering as an intractable part of the human condition" [9, p. 146].

her experience—is a natural extension of Marion's loving care of her patient, her family, and her self.[31] Her story shines a very bright light on that which medicine would keep hidden. Like all things locked away in dark corners of the mind, death is perverted by ignorance and denial, growing fearsome and strange. Once illuminated, as by Marion's story, death loses its fearsomeness, its power to appall. As we saw in Ashley's story, the denial of death is unhealthy: she had keenly suffered, in silence, for two years. Ashley could not respond to the imperative to care for herself in part because she had been trained to hide death away. To care for ourselves, we must face death honestly, without evasion: this is the only way to find meaning in our brief, mortal existence. As we see in Marion's story, caring for the dying can bring us to a fuller understanding of our own mortality and thus grant us access to the precious wealth of meaning in our life stories—if only we give ourselves the time and space to reflect. The stories the other members of the team shared, after hearing Marion read, were one and all stories of their own encounters with death. Taken together, it is as if these stories cry out with one voice, clear and strong, "Oh Death, where is thy sting?"[32]

Conclusion

This, then, is the conclusion of my ethics work with physicians and researchers over the past ten years: to honor the self-care imperative, one must attend to one's self-telling narratives. I therefore agree with David Morris that the "goal of narrative bioethics is to get the stories into the open, where we can examine their values, sift their conflicts, and explore their power to work on us" [15, p. 213]. Opening medical academicians to their own stories is crucial to faculty health. To effect this opening, the ethicist must contend with cultural and professional biases that work to keep these stories hidden. As Bruner writes, "The fact of the matter is that if you look at how people actually live their lives, they do a lot of things that prevent their seeing the narrative structures that characterize their lives. Mostly, they don't look, don't pause to look" [19, p. 8]. Fortunately, the native ability to see "the narrative structures" of our lives, while underused, is not rare. This ability, in fact, is almost certainly universal.[33]

I believe that the work of the biomedical ethicist must be devoted, in large part, to giving medical academicians opportunities to "pause to look," so that they might actualize their innate potential for narrative self-reflection. As Anne Hudson Jones argues,

---

[31] "The intersubjective certainty of death does not lose its terror, nor its countenance, but becomes familiar. Hence, the query of the other is always, on some level, about also this: and I, too, will die?" [35, p. 29].

[32] 1 Corinthians 15:55

[33] "Narrative capacity seems to be an innate human ability" [21, p. 160].

However natural a part of the human equipment, narrative skill can be developed by exposure to environments rich in stories. …One consequence of the increasing interest in narrative ethics is the need to think about the kind of training that can enhance the natural narrative capacities of those … whose professional training has not been focused on the study of literature or narrative. … Interpretation is a skill that can be learned and practiced by reading complex texts with the guidance of highly skilled readers and interpreters. [21, pp. 160–161]

Over the past ten years, a great deal of effort has been devoted worldwide to developing narrative training in the medical academy and to testing its effectiveness [21, pp. 163–164 and 15, p. 202]. At Columbia University Medical Center alone, the narrative medicine faculty has developed narrative training that includes literature and medicine reading groups, inpatient rounds, graduate seminars, writing groups, parallel charting curricula, intensive weekend workshops, and much more. Starting in fall of 2009, the Program in Narrative Medicine is offering a Master of Science in Narrative Medicine. This full-year Columbia University Master's program will provide intensive training at the university and the medical center for the theoretical, textual, clinical, pedagogic, and personal facets of a practice of and scholarly career in narrative medicine. Classes will gather a mix of health professionals and scholars from a variety of humanities and social science disciplines.

These narrative medicine initiatives are not focused primarily on care of the self. Indeed, they are, first and foremost, responses to the call of the suffering other, to the demand to offer her care [37]. Yet any such response is ultimately doomed unless it also answers that other call—the call to care for myself, to bring myself to light, and to tell and hear my own story. Like Ashley, I *must* hold Mary through her terror at death's approach. If I am not to lose myself to that terror, however, I *must* join my story—unique and whole—to the story we tell separately together: the always-unfolding story of why we tell ourselves stories.

References

1. Levinas, E. (1988) Useless suffering. Translated by Richard Cohen. In *The Provocation of Levinas*, edited by Robert Bernasconi and David Wood, 156–166. London: Routledge.
2. Cohen, R. (2001) *Ethics, Exegesis and Philosophy: Interpretation after Levinas* Cambridge: Cambridge University Press.
3. Boccher-Lattimore, D. (2008) Epidemiology. In *Faculty Health in Academic Medicine: Physicians, Scientists, and the Pressures of Success*, edited by Thomas R. Cole and Thelma Jean Goodrich, New Jersey: Humana.
4. Charon, R. (2006) *Narrative Medicine: Honoring the Stories of Illness* Oxford: Oxford University Press.
5. Dubose, E., Hamel R., and O'Connell, L. (eds) (1994) *A Matter of Principles? Ferment in U.S. Bioethics* Valley Forge, PA: Trinity Press International.
6. Nicholas, B., and Gillett, G. (1997) Doctors' stories, patients' stories: a narrative approach to teaching medical ethics *Journal of Medical Ethics* **23**, 295–299.
7. O'Toole, J. (1995) The story of ethics: narrative as a means for ethical understanding and action *JAMA* **273**, 1387–1390.
8. Poirier, S. (2002) Voice in the medical narrative. In *Stories Matter: The Role of Narrative in Medical Ethics*, edited by Rita Charon and Martha Montello, 48–58. New York: Routledge.

9. Frank, A. (1995) *The Wounded Storyteller: Body, Illness, and Ethics* Chicago, IL: The University of Chicago Press.

10. James, W. (1987) *Writings: 1902–1910* New York: Library of America.

11. Bosk, C. (1995) *All God's Mistakes: Genetic Counseling in a Pediatric Hospital* Chicago, IL: University of Chicago Press.

12. Nussbaum, M. (1986) *The Fragility of Goodness: Luck and Ethics in Greek Tragedy and Philosophy* Cambridge: Cambridge University Press.

13. Charon, R. and Montello, M. (2002) Memory and anticipation: the practice of narrative ethics. In *Stories Matter: The Role of Narrative in Medical Ethics*, edited by Rita Charon and Martha Montello, ix–xii. New York: Routledge.

14. Johnson, M. (1994) *Moral Imagination: Implications of Cognitive Science for Ethics* Chicago, IL: University of Chicago Press.

15. Morris, D. (2002) Narrative, ethics, and pain: thinking *with* stories. In *Stories Matter: The Role of Narrative in Medical Ethics*, edited by Rita Charon and Martha Montello, 196–218. New York: Routledge.

16. Martinez, R. (2002) Narrative understanding and methods in psychiatry and behavioral health. In *Stories Matter: The Role of Narrative in Medical Ethics*, edited by Rita Charon and Martha Montello, pp. 126–137. New York: Routledge.

17. Lindemann Nelson, H. (2002) Context: backward, sideways, and forward. In *Stories Matter: The Role of Narrative in Medical Ethics*, edited by Rita Charon and Martha Montello, pp. 39–47. New York: Routledge.

18. DasGupta, S. (2008) Narrative humility *The Lancet* **371**, 980–981.

19. Bruner, J. (2002) Narratives of human plight: a conversation with jerome bruner. In *Stories Matter: The Role of Narrative in Medical Ethics*, edited by Rita Charon and Martha Montello, pp. 3–9. New York: Routledge.

20. Brody, H. (2002) Narrative ethics and institutional impact. In *Stories Matter: The Role of Narrative in Medical Ethics*, edited by Rita Charon and Martha Montello, pp. 149–153. New York: Routledge.

21. Hudson Jones, A. (2002) The color of the wallpaper: training for narrative ethics. In *Stories Matter: The Role of Narrative in Medical Ethics*, edited by Rita Charon and Martha Montello, pp. 160–167. New York: Routledge.

22. Trautmann Banks, J. (2002) The story inside. In *Stories Matter: The Role of Narrative in Medical Ethics*, edited by Rita Charon and Martha Montello, pp. 219–226. New York: Routledge.

23. Richman, M. (2006) The fruit lady. Unpublished story given at The Center for Family and Community Medicine Narrative Medicine Inpatient Rounds on May 12, 2006. (Author requests that story not be reproduced without her express permission.)

24. Connelly, J. (2002) In the absence of narrative. In *Stories Matter: The Role of Narrative in Medical Ethics*, edited by Rita Charon and Martha Montello, pp. 138–146. New York: Routledge.

25. Burke, K. (1966) *Language as Symbolic Action: Essays on Life, Literature, and Method* Berkeley, CA: University of California Press.

26. Hoffmaster, B. (1994) The forms and limits of medical ethics *Social Science and Medicine* **39**, 1155–1164.

27. White, M., and Epston, D. (1990) *Narrative Means to Therapeutic Ends* New York: W.W. Norton.

28. Chambers, T., and Montgomery, K. (2002) Plot: framing contingency and choice in bioethics. In *Stories Matter: The Role of Narrative in Medical Ethics*, edited by Rita Charon and Martha Montello, pp. 77–84. New York: Routledge.

29. Carr, D. (1986) *Time, Narrative, and History* Bloomington, IN: Indiana University Press.

30. Charon, R. (2002) Time and ethics. In *Stories Matter: The Role of Narrative in Medical Ethics*, edited by Rita Charon and Martha Montello, pp. 59–68. New York: Routledge.

31. Anderson, C., and Montello, M. (2002) The reader's response and why it matters in biomedical ethics. In *Stories Matter: The Role of Narrative in Medical Ethics*, edited by Rita Charon and Martha Montello, pp. 85–94. New York: Routledge.

32. Forster, E.M. (1910) *Howards End* London: G.P. Putnam's and Sons.
33. Shem, S. (1991) Psychiatry and literature: a relational perspective *Literature and Medicine* **10**, 42–65.
34. Lukács, G. (1974) *Soul and Form*, trans. Anna Bostok. London: Merlin Press.
35. Hunsaker Hawkins, A. (2002) The idea of character. In *Stories Matter: The Role of Narrative in Medical Ethics*, edited by Rita Charon and Martha Montello, pp. 69–76. New York: Routledge.
36. Zoloth, L., and Charon, R. (2002) Like an open book: reliability, intersubjectivity, and textuality in bioethics. In *Stories Matter: The Role of Narrative in Medicine Ethics*, edited by Rita Charon and Martha Montello, pp. 21–36. New York: Routledge.
37. Irvine, C. (2005) The other side of silence: Levinas, medicine and literature *Literature and Medicine* **24**, 8–18.

Chapter 11
Faculty Health and the Crisis of Meaning: Humanistic Diagnosis and Treatment

Thomas Cole and Nathan Carlin

Abstract Recent concern for faculty health is a symptom of the damaged situation of contemporary health science centers; our human infrastructure is being compromised. This chapter argues that problems in faculty health often grow out of a crisis of meaning and identity that confronts health professionals increasingly unable to live up to their highest values and ideals. Recent trends in bioethics have emphasized concern for the patient as a whole person, but the patient is not the only whole person in the consulting room. Very little attention has been paid to the legitimate needs and concerns of the physician. Likewise, the research scientist is more than a machine for turning out grants and publications; yet the personhood of the researcher is almost nowhere acknowledged. This chapter offers perspectives from the humanities on faculty health. First, it sketches a historical context by locating faculty health within the recent crisis of academic health centers. Second, it uses the philosophical concept of "ethical violence" to cast new light on how academic physicians and scientists come to suffer and develop disease in striving to live up to their professional ideals. In conclusion, it discusses specific methods and programs in which the humanities and expressive arts provide avenues of reflection, release, and personal growth to replenish those whose work life requires, but virtually excludes these essential ingredients of meaning. Ideally, these programs will be part of a conscious commitment to compassion and care in institutional culture.

Keywords Academic health centers, history, health, personhood of physicians and scientists, humanities, ethical violence

On the plains of the upper Midwest, there was a time when farmers lost their lives during a blizzard. Early in the morning, they went to the barn to tend to the animals.

T. Cole
Director, John P. McGovern Center for Health, Humanities, and the Human Spirit, University of Texas-Houston Medical School, Houston, Texas, USA
e-mail: thomas.cole@uth.tmc.edu

N. Carlin
Rice University, Houston, Texas, USA

T.R. Cole et al. (eds.) *Faculty Health in Academic Medicine*,
© Humana Press, a part of Springer Science + Business Media, LLC 2009

When they came out to return to the farmhouse, they were blinded by a fresh snow-storm, lost their tracks, and froze to death in their own backyards. Over time, at the first sign of a blizzard, farmers learned to tie a rope between the house and the barn to find their way back home [1]. We live in a different yet no less-threatening blizzard today. We live and work at a time of crisis in academic health science centers, amidst a technocentric and dehumanized medicine, within a broken and unjust system of health care. And, to borrow a phrase from the Beatles, we are looking for ways to get back home.

Recent concern for faculty health is a symptom of the damaged situation of contemporary health science centers; our human infrastructure is being compromised. Our situation, of course, is neither unique nor isolated. Twenty-first century citizens in general are buffeted by the cold winds of globalization, the dizzying electronic pace of virtually everything, the vulnerability caused by a shrinking safety net, terrorism and warfare, and the alarming deterioration of our environment. All of us carry the weight of these trends to our life's work. Those who work in academic health centers face particular institutional strains caused by a marketplace restructuring of health care, a shrinking safety net, more indigent patients to care for, and declining federal support for research. Health, as an AAMC task force put it in 1999, "is not just the absence of disease but … includes a sense that life has purpose and meaning" [2, p. 24]. Health is a process through which individuals maintain this sense of coherence and their capacity to function in the face of internal and environmental changes [3]. Problems in faculty health, this chapter will argue, often grow out of a crisis of meaning and identity that confronts health professionals increasingly unable to live up to their highest values and ideals.

Recent trends in bioethics have emphasized concern for the patient as a whole person, but the patient is not the only whole person in the consulting room. The dehumanization of contemporary medicine affects faculty and caregivers, as well as patients and families. Very little attention has been paid to the legitimate needs and concerns of the physician. Likewise, the research scientist is more than a machine for turning out grants and publications; yet the personhood of the researcher is almost nowhere acknowledged. A blizzard seems to separate daily working lives from the purposes and meanings of academic medicine.

Our chapter is rooted in the medical humanities, an interdisciplinary field which has grown up in the last 35 years to address fundamental human issues generated by new scientific knowledge and technological capability. Beginning in the late 1960s, and influenced by the atrocities of World War II and the Nuremberg trials, leading physicians, theologians, lawyers, and philosophers realized that scientific medicine—a profession that had explicitly detached itself from broader frameworks of meaning and value—was not intellectually equipped to handle the moral and existential questions produced by its own power [4]. As a result, entirely new fields of academic inquiry, education, and professional practice—known as bioethics and the medical humanities—arose to grapple with problematic issues such as the protection of research subjects, the goals of medicine, the definitions of death, the rights of patients, the cessation of treatment, the meaning of illness, and the distribution of health care resources.

The humanities complement scientific knowledge by addressing moral, aesthetic, and spiritual issues, and by articulating their historical and cultural contexts. But what are the humanities? And why do they matter in contemporary medicine and science? The humanities can be defined in terms of disciplines, subject matter, or methods. Defined by disciplines, the humanities range from languages and literature, history, and philosophy to religious studies, jurisprudence, and those aspects of the social sciences that emphasize interpreting, valuing, and self-knowing. Defined by its subject matter, the humanities reflect on the fundamental question: "What does it mean to be human?" [5]. As the Rockefeller Commission on the Humanities put it in 1980: "[the humanities] reveal how people have tried to make moral, spiritual, and intellectual sense of a world in which irrationality, despair, loneliness, and death are as conspicuous as birth, friendship, hope, and reason. We learn how individuals or societies define the moral life and try to attain it, attempt to reconcile freedom and the responsibilities of citizenship, and express themselves artistically" [6, p. 1]. From the perspective of methods, the humanities have been defined as the proper cultivation of the four essential "arts": language, analysis of ideas, literary and artistic criticism, and historiography [7]. Rather than mathematical proof or reproducible results, humanities research and teaching are dedicated to understanding human experience through the disciplined development of insight, perspective, critical understanding, discernment, and creativity.

But disciplines, subject matter, and methods—whether taken separately or together—cannot adequately characterize the humanities, which ultimately emphasize description, interpretation, explanation, and appreciation of the variety, uniqueness, complexity, originality, and unpredictability of human beings striving to live and understand themselves. I (Thomas Cole) direct the McGovern Center for Health, Humanities, and the Human Spirit at the University of Texas Health Science Center, Houston. As such, I work daily with faculty and colleagues to provide medical education, research, and change interventions that cultivate humanistic ways of knowing and caring in health care.

In particular, I am inspired by the ancient educational ideal originally known as *humanitas*—whose original Latin meaning was "human feeling." Gradually, the word *humanitas* became associated with a holistic educational ethos that blended knowledge, humane feeling, and action. We aim to educate emotionally and spiritually integrated individuals who exemplify this wonderful and elusive mix of knowledge, feeling, and action. There is no method or protocol for creating such integrity. And the aim is not simply to bring humanities content to the students and faculty. The aim, rather, is to evoke and nurture their humanity through dialogue and reflective engagement in small group settings. We encourage the lifelong pursuit of self-development in the service of others. While self-sacrifice is often called for in the course of caring for others, our focus on concern for one's own health and well-being is no small aspect of our educational ideal.

This chapter offers perspectives from the humanities on faculty health. First, it sketches a historical context by locating faculty health within the recent crisis of academic health centers. Second, it uses the philosophical concept of "ethical violence" to cast new light on how academic physicians and scientists come to suffer and develop disease in striving to live up to their professional ideals. In conclusion, we discuss specific methods and programs in which the humanities and expressive arts provide avenues of reflection, release, and personal growth to replenish those whose work life requires but virtually excludes these essential ingredients of meaning. Ideally, these programs will be part of a conscious commitment to compassion and care in institutional culture.

The Recent History of Academic Health Centers

While there is no universally accepted definition of an academic health center (AHC), there is general agreement that these centers, as we know them today, came about during the second-half of the 20th century [8]. To be sure, they had their roots in previous decades, e.g., in the Flexner Report of 1910, the fundamental modern curricular reform in medical education, and in the increase of federal funding for medical research during and after World War II. In any case, in the 1960s, the establishment of Medicare and Medicaid, which provided government-supported health care for the elderly and the poor, created vast new sources of clinical revenue for medical schools, which in turn funded the expansion of collaborating institutions.

As William Rothstein notes in *American Medical Schools and the Practice of Medicine*, medical schools often occupied contiguous geographic space with other health occupations. Hence academic health centers came into existence through collaboration among institutions in the same area. Therefore, academic health centers have usually "consisted of a medical school, a hospital, and at least one other professional school" [9, p. 225]. One reason why it is difficult to define what precisely constitutes an academic health center is that, as Rothstein writes, academic health centers "have varied in their relationship to the parent university and in their organizational structure" [9, p. 226]. The creation of an academic health center, then, is somewhat of an organic process, which is why it is sometimes said, "If you have seen one academic health center, you have seen one academic health center."

It is well known that in recent years academic health centers in the United States have been and are going through a period of transition [10–13]. This transition is deeply political and rooted in market forces, which have removed the financial privileges that AHCs once enjoyed, thereby putting new pressures on research, education, and clinical care. Medical school faculty, previously given funded time for teaching and research, are increasingly drafted to bring in clinical revenues to cover their salaries.

In 1993, in an issue of *The Western Journal of Medicine* examining the relationship between government and academic health science centers in the United States, Lloyd Smith noted that it was no accident that the American Association of Medical Colleges had recently moved its headquarters from Chicago to Washington, DC. Something had changed in the culture of medical education and research in the

1990s. The American Association moved to Washington because they realized that, as Smith puts it, "The groves of academe, at least in medicine, are no longer tranquil sites for detached scholarship. The imperatives of public policy increasingly intrude, seduce, and command, and the pace of these incursions is accelerating" [14, p. 211]. In recent articles about academic health science centers, there is often a brief section giving the history of these centers, and, inevitably, they note that these centers are in financial trouble today and that this is, in part, due to the fact that these centers no longer enjoy the amount of federal funding they did in previous years [15]. So in the early history of academic health centers, the story is told of how supportive and instrumental the government was in the formation of academic health centers, but now the tone is different—now academic health centers need to justify themselves and their work.

When did this change occur? Cornelius Hopper and Cathryn Nation argue that this shift occurred in the 1990s, and that it was a response to the "crisis in the American health care system at the turn of a new century" [13, p. 144]. The changing tide of AHC finances was "the culmination of three decades of experiments, mostly unsuccessful, designed to control the spiraling costs of health care while also providing a reasonable level of access to care for all segments of the American population" [13, p. 144]. In the 1970s and the 1980s, Hopper and Nation point out, there were various attempts at reform, such as the certificate of need and federally sponsored health insurance programs for the aged and poor. But in the 1990s, there was a shift away from fee-for-service toward a system of capitation. "No institution has been more negatively affected by this revolution," the authors suggest, "than the academic medical center" [13, p. 144].

If Hopper and Nation are correct, academic health centers began to lose funding sometime in the early- to mid-1990s. Around the time this shift was taking place, one influential commentator on the status of academic health centers, David Blumenthal, conducted a study of seven of these centers [16]. David Blumenthal and Gregg Meyer visited seven academic health centers for two days each, conducted interviews with senior managers and clinicians, and reviewed internal papers. They chose their sites on the basis of membership in the Association of Academic Health Centers (AAHC), which, at the time, consisted of "126 institutions that combine professional schools with teaching hospitals and have simultaneous commitments to teaching, research, and clinical care" [16, p. 201]. Blumenthal and Meyer realized that their findings would not be representative, but they nevertheless "sought to include institutions that varied in ownership (public and private) and geographic setting (urban and rural)" [16, p. 201]. "Although our resulting sample shows considerable diversity," they write "it may somewhat overrepresent large, eminent AHCs in markets that are undergoing competitive transformation" [16, p. 201].

In "Academic Health Centers in a Changing Environment," Blumenthal and Meyer report the results of their 1994 study. They found that "the central problem facing AHCs relates to the effect of market restructuring on their revenue sources" [16, p. 202]. They also noted that academic health centers provide two kinds of goods, public and private. Public goods are those services rendered such as teaching, research, and clinical services to society at large, while private goods are those services rendered for which there is a private market from individuals or organizations.

The problem was that market forces were forcing these centers to charge less for their private services, while the political climate resulted in the reduction of government subsidies for public services [16, p. 204]. Blumenthal and Meyer made these recommendations in light of the financial stresses:

1. Increase sales of clinical services to private purchasers
2. Reduce costs of clinical services
3. Increase sales of clinical services to government purchasers
4. Increase private investment supporting biomedical research
5. Increase sales of nonclinical services and teaching materials to external markets
6. Reduce costs of nonclinical services, i.e., teaching and research

These strategies, all of which are pursued today, make good sense in terms of balancing the budget. They result in cutbacks that reduce funding for teaching and research while increasing financial pressure on physicians, scientists, and other health care providers. These strategies may succeed, but they do not help us find our way in the blizzard. Financial success may go hand in hand with human failure. Spreadsheets do not help institutions find their moral bearings.

Ethical Violence

In *Problems of Moral Philosophy*, the German refugee social theorist Theodor Adorno called attention to a cruel aspect of collectively enforced morality [17]. In Adorno's view, any set of ethical maxims or rules must be appropriable by individuals "in a living way." When an ethical norm "turns out, within existing social conditions, to be impossible to appropriate," the result is ethical violence. Institutions that ignore existing social conditions and rigidly enforce moral rules are actually perpetrating violence on those expected to do the impossible [18]. The more desirable course of action is for institutions to acknowledge the gap and to develop a mutual adjustment between the norms and existing social and cultural conditions.

While Adorno had political history in mind, particularly cases in which the powerful enforce their norms on minorities in the name of universal morality, his analysis of ethical violence can help us understand problems of faculty health in a new way. Our institutions are filled with people of faith, good will, and commitment who strive to "do the right thing," to:

• Provide compassionate and ethically sound care
• Teach and mentor students
• Maintain scientific standards of practice
• Keep current with the most recent literature in one's field
• Undertake important and innovative biomedical research and
• Bring in revenues through grants or clinical work.

These are all good and necessary activities, to be sure, activities rooted in professional ideals of science, service, education, and spiritual development, and they are widely taught and internalized by health care professionals. Yet current conditions severely limit the ability of faculty to live up to these requirements and ideals, which creates a cognitive dissonance that we believe leads to cynicism, disillusionment, self-doubt, disease, and a retreat from ideals that seem so obviously unrealistic. The term burnout is a euphemistic shorthand for a deeply troubling process. As Maslach and Leiter put it, "Burnout is the index of dislocation between what people are and what they have to do. It represents an erosion in values, dignity, spirit, and will—and erosion of the human soul. It is a malady that spreads gradually and continuously over time, putting people into a downward spiral from which it's hard to recover" [19, p. 17].

Another dimension of ethical violence might be called political violence. Our country is founded on the idea that the exercise of arbitrary power is injurious to public health and well-being. Academic health centers, of course, will never be model democracies; but we must be mindful of the demoralizing, disrespectful, and destructive consequences of the abuse of power. What constitutes the abuse of power? Power exercised hierarchically and impersonally, without explicit recognition of all relevant stakeholders, without consultation, and the opportunity for dialogue. Contemporary medicine recognizes the ethical principal of "respect for persons" in research and patient care. But academic health centers do not always show respect for the personhood of their faculty and employees.

A recent conversation with a very successful colleague here at UT Houston School of Medicine exemplifies this problem. At a faculty meeting his Chair pulled out a spreadsheet of salaries and clinical income, and simply informed them that they had to increase clinical revenues at the risk of reduction in salaries. Such conversations take place all over the country, and there is nothing new or necessarily wrong with this policy. In fact, clinical revenues often provide the margin to subsidize important non-revenue producing activities such as teaching and research.

What my colleague found demoralizing was the absence of any reaffirmation of the mission of the department in the first place. From the chair's actions and conversation, there was no way of distinguishing the functioning of an academic clinical department from the functioning of a private group practice—except that individual income would be higher in the private sector. Nor was there any dialogue or statement of concern about the stress imposed on individual faculty members. The point here is not to criticize merits of the policy (a different issue), but to critique an inadequate leadership practice which relies on arbitrary power.

So our argument here is that the compromised health of our faculty emerges in part from the ethical violence and unreflective hierarchy inherent in our institutional life.

This is not to blame institutional leadership, however. Institutional leaders clearly are not in control of the external conditions that distort, threaten, and blight academic health science centers. But they do make basic resource allocation decisions, often without discussion or a clear rationale. The key point is that the physical, moral, and spiritual health of faculty cannot be addressed *without* the commitment of institutional leaders, and this commitment should include genuine consultation, communication, and concern. This means conveying a real and tangible commitment to the well-being of faculty who need to be supported, nurtured, acknowledged for their struggles, and made aware of the need for self-care so they can more reasonably put on the Aesculapian mantle and flourish in their mission. At the same time, individual faculty members have a responsibility for self-care. No amount of complaining about institutional callousness and lack of support can substitute for individuals personally grappling with their own priorities, their own physical, emotional, and spiritual well-being. Institutions may be responsible for providing the rope, but individuals must anchor themselves and find their own way back home.

Conclusion: The Integrity That Comes from Being What You Are

How can the medical humanities attend to the needs and concerns of faculty? Essentially, by facilitating active engagement with relevant literature, history, ethics, the arts, and spiritual/religious resources. The humanities provide faculty with opportunities to reflect, replenish, and renew themselves, to gain perspective, insight, and understanding of critical issues in health and health care. Lectures or seminars, for example, on topics such as the ethics of health policy, history of medicine, the experience of suffering, the nature of healing, reproductive ethics, and so forth break the daily routine and help individuals address aspects of medicine and scientific research where technical mastery is impossible, ethical problems are difficult, and existential meaning is hard to come by.

In our highly visual culture, documentary and dramatic films are an increasingly common and effective way to stimulate deep engagement. In my documentary film *Still Life: The Humanity of Anatomy*, for example, a dramatic dialogue between students in the anatomy lab and individuals who will donate their bodies encourages viewers to ponder their own mortality and to think about the future of their own bodies after death. Michael Moore's current film *Sicko* dramatizes the deepening crisis of the health care system we work in every day, as well as the political opportunities for reform. Films such as *Ikiru, Million Dollar Baby, The Cider House Rules, Wit, Gattaca, and Tell Me a Riddle* are a only tiny fraction of the available films.

Workshops and retreats provide another important resource for those faculty willing to risk opening themselves emotionally. Rachel Naomi Remen, for example, has developed a valuable series of workshops about Meaning in Medicine, modeled after her "Healer's Art" course, which is widely used in undergraduate medical education. Another important form of emotional and spiritual renewal in medicine has recently grown out of the work of Parker Palmer. Palmer, whose work began in the reform of secondary education [20], has recently moved into law and the health professions. His programs consist of a series of retreats and workshops based loosely on Quaker spirituality. In 2002, Dr. David Leach, Director of the American Council of Graduate Medical Education, created the Parker J. Palmer "Courage to Teach" Award, given annually to the directors of ten medical residency programs that exemplify patient-centered professionalism in medical education. Leach encourages residency directors to participate in the Parker Palmer retreats and workshops and to integrate aspects of them into residency education.

Palmer's programs are based on the idea that contemporary professional life creates an inner division within individuals: the "soul" is disconnected from the "role." His workshops and retreats use the model of a "circle of trust," in which various exercises (e.g., reflexive writing, meditation) and materials (e.g., poetry, music) are used to honor and encourage each member of the circle to overcome this division through a process of spiritual formation. University of Texas-Houston and University of Texas M. D. Anderson have recently collaborated in sponsoring a series of faculty renewal groups modeled loosely on Palmer's work. After two years of leading these faculty renewal groups, Henry Strobel, myself, and Thelma Jean Goodrich found that a small core group of participants reported a rich and rewarding experience.

Seminars and workshops in narrative medicine (or personal writing) provide another important avenue of self-expression for faculty, residents, and students. In small group settings—perhaps in a hospital conference room or at a faculty meeting—participants are asked to write quickly and without editing for 5–7 minutes. They write in response to prompts such as: write about the suffering of a patient who has moved you; write about a difficult patient from the patient's point of view; write about the first patient who died under your care; write about the most gratifying patient you have cared for; and so forth. Participants then read their writing out loud and others in the group respond. In my experience, these groups are uniformly valuable and often powerful. And there is very strong evidence that similar groups have positive personal results. We have yet to study the outcomes of these groups among medical faculty, research scientists, and students. From experience, we expect that findings will include grieving unmourned losses, reduced sense of isolation, reduced depression, and healing from painful experiences—both through the writing itself and the support of other group members [21]. In addition, writing groups promote narrative competence [22] through more careful listening to patient stories—often an important corrective to premature diagnosis and/or stereotypical thinking that lead to physician errors [23, 24].

The purpose of this essay has been twofold: (1) to analyze the crisis of faculty health in socio-historical, cultural, and philosophical terms; and (2) to point out that the humanities and the arts provide intellectual and spiritual experiences and texts that are interpreted, engaged, and pursued through dialogue. We have argued that the humanities and arts offer individuals an opportunity to find their moral bearings in the context of a community of learners, each seeking the integrity that comes from being who they really are. At their most effective, humanistic programs will be one element in an institutional culture committed to compassion and care for faculty and employees, as well as for patients [25].

References

1. Palmer, P. (2004) *A Hidden Wholeness: The Journey Toward an Undivided Life* San Francisco, CA: Jossey-Bass.
2. Medical Schools Objectives Project (1999) *Contemporary Issues in Medicine: Communication in Medicine, Report III* Washington, DC: The Association of American Medical Colleges.
3. Antonovsky, A. (1987) *Unraveling the Mystery of Health: How People Manage Stress and Stay Well* San Francisco, CA: Jossey-Bass.
4. Carson, R. (forthcoming). Engaged humanities: Moral work in the precincts of medicine.
5. *President's Council on Bioethics: Being Human* (2003) Washington, DC: GPO.
6. *The Humanities in American Life: Report of the Rockefeller Commission on the Humanities* (1980) Berkeley, CA: University of California Press.
7. Crane, R.S. (1967) *The Idea of the Humanities and Other Essays* Chicago, IL: Chicago University Press.
8. Kohn, L. (ed.). (2004) *Academic Health Centers: Leading Change in the 21st Century* Washington, DC: National Academies Press.
9. Rothstein, W. (1987) *American Medical Schools and the Practice of Medicine: A History* New York: Oxford University Press.
10. Crichlow, R. (1996) The evolution of a contemporary academic health care system *Archives of Surgery* **131**, 237–241.
11. Weissman, J., Saglam, D., Campbell, E., Causino, N., and Blumenthal, D. (1999) Market forces and unsponsored research in academic health centers *Journal of the American Medical Association* **281(12)**, 1093–1098.
12. DeAngelis, C. (2000) The plight of academic health centers *Journal of the American Medical Association* **283**, 2438–2439.
13. Hopper, C., and Nation, C. (2001) The academic health center in transition: overview *Archives of Surgery* **136**, 144–146.
14. Smith, L. (1993) Government and academic health science centers *The Western Journal of Medicine* **159(2)**, 211–212.
15. Johns, M. (1996) A perspective on the history of the academic health center *Laryngoscope* **106**, 1059–1062.
16. Blumenthal, D., and Meyer, G. (1994) Academic health centers in a changing environment *Health Affairs* **15(2)**, 201–215.
17. Adorno, T. (1963/2001) *Problems of Moral Philosophy* Stanford, CA: Stanford University Press.
18. Butler, J. (2005) *Giving an Account of Oneself* New York: Fordham University Press.
19. Maslach, C., and Leiter, M. (1997) *The Truth About Burnout: How Organizations Cause Personal Stress and What to do About it* San Francisco, CA: Jossey-Bass.
20. Palmer, P. (2007) *The Courage to Teach* San Francisco, CA: Jossey-Bass.
21. Anderson, C.M., and MacCurdy, M.M. (2000) *Writing and Healing* Urbana: National Council of Teachers of English.
22. Lepore, S.J., and Smyth, J.M. (2002) *The Writing Cure* Washington, DC: American Psychological Association Press.
23. Charon, R. (2006) *Narrative Medicine* New York: Oxford University Press.
24. Groopman, J. (2007) *How Doctors Think* Boston, MA: Houghton Mifflin.
25. Grigsby, K. (2007) Organizational culture and its consequences. Presented at the faculty health conference, University of Texas Health Science Center, July 19–21.

Chapter 12
Retaining and Reclaiming the Call of Medicine

Henry W. Strobel

Abstract Medicine has been understood as a calling from the earliest memory of humankind. There was always something unique, something out of the ordinary, that characterized and characterizes still the healer—whether shaman, curandero, herbalist, or physician/surgeon. This special calling in early times was recognized by the community generally to be based on some signal demonstration of the power for healing or some unique instance of a natural phenomenon. In later times an apprentice system arose in which the calling was passed from parent to child or from master to junior colleague. Later still, the recognition of a teachable skill led to the elaboration of requirements for licensure of the practice of healing in each of the healing arts. Nonetheless the notion of the calling remains though changed by time and the march of knowledge (scientia/science). For instance, the Hippocratic Oath (Table 12.1) is taken by each year's graduating class. Although the oath has been sanitized by the elimination of swearing by Apollo and the elimination of certain other phrases, it preserves the notion that the practice of medicine is still a calling, a vocation rather than a license alone.

Keywords Calling, vocation, values, oath, dedication

The notion of a calling culminating in taking an oath provides the touchstone for renewing and refreshing the dedication to medicine and especially to academic medicine. Through all challenges and difficulties, the promises in the oath remain. In remembering and renewing those promises lies the strength to continue in the academic roles of teaching, research, and modeling medical care. Contact with students and trainees calls to mind those promises once again and at the same time brings hope for the future. With hope comes renewed enthusiasm for the tasks. When all other reasons for continuing seem to dim, the promises remain and the promises lived out in our lives bring renewal with their remembrance.

H.W. Strobel
Associate Dean Faculty Affairs, Office of Faculty Affairs, Professor, Department of Biochemistry and Molecular Biology, The University of Texas Medical School, Houston, Texas, USA
e-mail: henry.w.strobel@uth.tmc.edu

T.R. Cole et al. (eds.) *Faculty Health in Academic Medicine,*
© Humana Press, a part of Springer Science + Business Media, LLC 2009

Two years ago my son, Nathaniel, joined the faculty of the University of Texas Medical School as Assistant Professor of Pediatrics in Critical Care. I asked him how he was enjoying his life as a faculty member in the department and medical school in which he had completed his M.D. degree, his residency, and his specialty fellowship. His response was, "I love what I do. I feel each day that I am on service that I make a difference in the life and the future of my patients. I love teaching fellows, residents, and students that they can also make a difference in lives and how they can make that difference in Critical Care." This answer was gratifying to me in all the ways that a parent is pleased when one's son or daughter discovers and claims the passion that drives him or her. The answer was especially gratifying to me because it started with and recognized the centrality of love in one's dedication to the field of choice, one's vocation or calling. It demonstrates the connection between "I love what I do" and "I cannot not do this". In this connection lies the deep meaning of profession/vocation.

Medicine is recognized as a vocation, a profession, and a learned calling. These three descriptions share a similar core meaning, which gives medicine (and other similar pursuits) a distinction from other occupations, That is, medicine connotes work in service of and for others. In fact, the Latin root meaning of vocation is "to call out" (the perfect passive infinitive *vocari* means to be called). Profession derives from the Latin base *fateri* (to own/to claim) and *pro* (on behalf of; for). This focus on the well-being of others is a hallmark of what comes to mind when the word medicine is used. To be sure, many other images are called up as well, but the focus on the other in medicine is its clearest and oldest base. A clever distinction between an occupation/job and profession is found in the quote "A job is what one does to live, but a profession is what one lives to do". The quote highlights the idea that the calling to a profession is something one cannot refuse and cannot refrain from doing.

Medicine as a profession or vocation calls persons by and to its nature, its practice, and its involvement in the lives of others. What is the response to this call? Clearly, for those who respond to this call, the answer is yes. This affirmative answer, however, is multilayered. First and fundamentally, there is the yes to the call itself—the claim of and for the role of physician. At a second level, however, one says yes to the standards of the field of medicine by assenting to its rules and requirements. At a third level, one assents to the other-directed goals and aims of the practice of medicine. At a fourth level, one accepts the challenge to the personal commitments inherent in dedication of the self to the well-being of others [1–3].

These layers of assent are, in general, temporally associated. The first yes represents the point of decision wherein one falls in love with the idea of medicine, with what appears at the moment of decision to be the perfection of fit between one's life and one's view of medicine. The second layer of assent often occurs in finding fulfillment in the mastery of the material of medicine—the ways and means of the practice of medicine. The third layer is a public subscription to the commitments/rules/ethics of practice and is imaged most clearly at the taking of the Oath of Hippocrates (Table 12.1). The fourth layer is wider in its temporal range. It is a private and personal embrace, an inculcation, of the values of medicine within one's

Table 12.1 The Oath of Hippocrates

The Oath of Hippocrates
I do solemnly swear, by whatever I hold most sacred:
That I will be loyal to the profession of medicine and just and generous to its members;
That I will lead my life and practice my profession in uprightness and honor;
That into whatsoever house I shall enter, it shall be for the good of the sick to the utmost of my power, holding myself far aloof from wrong, from corruption, from the tempting of others to vice;
That I will exercise my profession solely for the cure of my patients, and will give no drug, perform no operation, for a criminal purpose, even if solicited, far less suggest it;
That whatsoever I shall see or hear of the lives of men which is not fitting to be spoken, I will keep inviolably secret. These things do I swear.
And now, should I be true to this, my oath, may prosperity and good repute be ever mine; the opposite, should I prove myself forsworn

self. It is ultimately a "falling in love with" the individuals served through the skills, the knowledge, and the wisdom acquired both in training and in practice. These "yeses", these layers of assent, are to the profession of the physician the ways of claiming the role of the physician, and the proclaiming of their reality by one's own actions and beliefs.

Of these layers of assent the most publicly known and visible is the Oath of Hippocrates (Table 12.1). The Oath is a highlight and is most often the crowning event of medical school commencements. It is the public promise made by the new physicians to live publicly according to their highest aspirations and to the expectations of the community of physicians. The very public nature of the promises in the Oath intensifies the desire to live into the "yeses" of the call to medicine. Therefore, the Oath serves not only as the visible sign of a contract made with the community, but also as an invisible and personal assumption of a new role in which the new physician will live life.

Research physicians and research scientists who are part of academic medical centers live out these dedications in a different way. They focus on increasing knowledge and technologies for the welfare of patients although they do not necessarily care for patients. Nonetheless, the same intensity and commitment is present. For research scientists, however, there is not an equivalent to the Oath of Hippocrates which is taken in public. Commitment and intensity are privately maintained. Perhaps an analogous oath would be helpful to underscore that all faculty do indeed work for the public and in the public eye.

I also have experienced this process of oath-taking as the culmination of an educational experience with its examinations and certifications. I was ordained to the Priesthood in 1979 in a very solemn service presided over by the Bishop of the Diocese in the presence of many witnesses. By my answer to a number of questions before witnesses, I professed obedience to a series of rules, practices, and beliefs. Then amidst a great cloud of incense, the Bishop intoned the Veni Creator Spiritus over me. Finally, he said the ordination prayer through which I became a priest and joined the company of other priests. That solemn and unforgettable process gave me a new role. I gained faculties in the process. I was enabled to consecrate the bread and wine during the Mass. I was enabled to pronounce absolution of sins in public and private confession. I was enabled to bless marriages. However, at my ordination I also lost faculties. I lost the capacity to be a member of any parish, to hold any parish office, to vote in any parish election. There were other losses as well. To some considerable extent, I lost autonomy. I was subject to the will of others in terms of where and how I practiced this new profession. I also to some extent lost my privacy by becoming a person for others through the vows I had taken. Thus, I draw the conclusion that by profession and in any profession, one gains abilities/license/faculties and yet with that gain there are also losses.

When my son was graduated from medical school in 1992, he was a member of a large class who together professed their ascription and obedience to the rules and standards governing the practice of medicine. Making the promises of the Hippocratic Oath together was a culminating experience for an already close class. The corporate nature of the oath-taking holds the members of the class together as they practice medicine in different areas and states. Yet my son also expresses loss amidst the gains of being able to practice medicine. He has less time to spend with his family [4]. His schedule is not his own to make. When he commits to an event or meeting, it is with the knowledge that he may not be able to keep that commitment because of his prior commitment to medicine made in the words of the Hippocratic Oath. He loves what he does. He cannot not do it in spite of the loss he sees and knows. In this I believe my son is no different from any other physician.

If physicians love what they do as physicians—teach, do research, take care of patients, if research scientists and physician scientists love what they do, if they do what they do because they cannot not do it, why then is there disaffection with being a physician or a research scientist in an academic medicine center?

Part of the challenge facing academic medical centers derives from the unintended consequences of well-intentioned changes in medical practice, health care delivery, and opportunities/funding for research. While the intentions are all but universally applauded, the unintended consequences of these changes have led to increased pressure on physicians, particularly those in academic medical centers, where teaching and scholarly activities are important criteria for promotion in academic rank. Table 12.2 lists several illustrations of this connection between well-intended changes and unintended consequences. For example, one intention is to make medical care more conveniently accessible to patients. Therefore, many academic medical centers have opened additional sites in surrounding and distant neighborhoods. While this change does improve access and in many cases broadens the payer mix for academic medical centers, it has brought about an unintended sorting of faculty members within a department among the ancillary sites, thereby making it harder for faculty to know one another or collaborate with one another and participate in departmental and center-wide committees. Often this separation leads to a sense of "second class" status if one is assigned to a distant site rather than the primary site.

Financial challenges affect physicians in academic medical centers as well as those in private practice groups and those who practice as hospitalists. They seem to affect academic medical centers more acutely than others because academic

Table 12.2 Intentions to improve medical care and their unintended consequences

Intention	Unintended consequence
Make medicine more available to all	Unreimbursed or under-reimbursed medical care costs rise and income decreases, thereby making some sources and some venues unsustainable
Make medical care more accessible to patients by opening satellite sites	Faculty assigned to satellite sites are disconnected from the medical center with less interaction, less collaboration, and less scholarly activity
Eighty hour work week for residents	Faculty spend more time on call and do not receive remuneration for it, a situation that increases the number leaving [5]
Reduce the costs of medical care	Pressure to support services where procedures are billable only at rates less than the cost of service increases deficits and leads to reductions of positions, thereby increasing the load on the remaining staff

medical centers are very often sites of care for the less well-insured and for those with no insurance at all. It may be argued that the inefficiencies due to teaching students may also contribute to this problem since private university medical school-associated hospitals do the same teaching and do not seem to suffer the consequences at the same level. All of the challenges facing current medicine affect the busy physician at some level, but this is especially the case for the faculty of the academic medical center.

An analogous set of intentions/unintended consequences exist for research scientists and physician scientists in an academic medical center. Again the intentions serve useful goals and are recognized as sensible managerial approaches. Table 12.3 lists some of these intentions and their unintended consequences.

In order to provide a system for evaluating faculty on this research and teaching commitments and accomplishments in very different venues, many centers have adopted and adapted for specific purposes a relative value unit (RVU) system. The intent of this system is to assign relative values to teaching efforts in standard lecture formats as well as in other formats such as problem-based learning, team-based learning, laboratory, clerkship, selective, small group, fourth-year elective, and ambulatory medicine selectives [6, 7]. Some systems give various RVU credits for membership on a graduate student committee and for being a committee chair. The unintended consequences derive from differing values for the same time spent teaching in differing venues and from lack of value placed on service to community responsibilities such as membership on university committees for promotion, faculty senate, or admissions. Moreover, while a member on an National Institutes of Health (NIH) study section or a review panelist for any other granting agency is reimbursed mostly for costs, and while these appointments carry honor, importance on the CV and impact for the promotions committee, no RVU credit recognizes the value of these critical community functions. This oversight seriously undercuts

Table 12.3 Intentions to improve faculty/research effectiveness and their unintended consequences

Intention	Unintended consequence
The RVU system is used to evaluate teaching and research activities of faculty	Service on university committees undervalued or not rated as RVU-worthy leads to reduced interest in activities that benefit the university or wider community, e.g., admissions committee, NIH study sections
The organization of NIH study section changes to reflect national priorities along organ system lines	The study section that reviewed the previous grant no longer exists as situation forcing A1 and A2 applications and leading to interruption of funding and loss of staff
NIH Roadmap emphasizes Conquest of Disease targets	Large groups are favored for efficiency and the number of individual investigator single grant laboratories is reduced
Risk management and compliance guidelines are implemented without faculty involvement	The sense of collegiality between faculty and system administrators diminished and the perception of barriers to research increases

voluntary activities inside and outside of the institution and has an overall negative effect on interest in research, obtaining grant funding, and scholarly activities in general as unintended consequences.

The NIH has altered and is altering again the structure and charges to panels for the study sections. The overall goal is to have the restructured panels organized along organ systems lines with a focus on organ-related disease processes. This realignment is consistent with national priorities on organ system diseases. Many previously extant study sections have been eliminated and/or folded into new amalgamated review panels in this process. The unintended consequences include the reality that the study section which previously reviewed one's grant is no longer extant and has been replaced by one with few if any names recognizable for their expertise in the topic area of the grant proposal that one has just submitted. It is a not an infrequent experience that grants must go through A1 and A2 iterations before the study section is able to rate the proposal with a score sufficient for funding. The interruption is often an interruption of productivity or disruption of skilled research teams.

The NIH Roadmap has a laudably broad view, which emphasizes Conquest of Disease targets. An unintended consequence seems to be increased difficulty for the single grant laboratory especially that of the new investigator or new-to-this-field researcher. Consequently, there may be an unintended preference for large-group laboratories with multiple grants over the smaller laboratory with one or two grants.

The intention to implement risk management and compliance guidelines arises in response to federal, state, and local mandates/concerns. Most often these are "top-down" directives that seem to the faculty to arise from some "Dark Tower" "over there." The absence of communication is the major issue and leads to a break in the sense of collegiality between faculty and system administrators. The increasing list of regulations appears to many to be a barrier to research and scholarly activity and therefore gives rise to a sense of ennui and discouragement.

Now that the challenges for faculty in academic medical centers have been delineated, the next task is to identify the effects of these challenges on faculty. The simplest way to name the primary result of these challenges on some faculty is *accidia*—the loss of heart. Its sources are many. The first source is the combination of circumstances, which inhibit one's ability to give one's gift to others. The causes could be time pressure, patient overload, lack of sufficient staff support, etc. All of these combine to restrict the conventional interaction between patient and physician, between academician and administrator. Further, dealing with the depressed and underappreciated attitudes of other health care delivery colleagues and academicians increases the loss of heart, as well, there is a loss of hope for improvement in providing caring treatment for patients and a supportive environment for teaching and learning. The cost-focused behavior of superiors complicates health care delivery and scholarly activity by reducing the number of personnel, increasing the work-load, and thereby compressing the time available for each patient and each trainee. These features and other issues of salary level, billing, collection levels, compliance, and charting inefficiencies lead to loss of heart in faculty and loss of faculty from institutions.

What can be done to counter loss of heart? While there are many institutional changes that can and probably should be made, perhaps there are changes an individual can make. In connection with similar changes in others, these can bring about a change in the working environment of the institution.

(a) It is important to take joy in relating to each patient and each student as a human encounter that sustains the covenant between physician and patient and between teacher and student.
(b) It is important to remember the passion that led to the decision to do medicine, to become a scholar. Perhaps that altruism expressed in the essay applying to medical school still resonates with heart, spirit, and experience in more ways than when it was written. Let recollection of the process and pathway to the present place give light and hope to the discouragement in today's trials.
(c) It is important to rejoice in verbal and written comments from patients, students, and colleagues. Perhaps there is no truer prompt to do one's utmost than to hear appreciation or admiration expressed for past work.
(d) It is important to remember that being an academician/being a physician is a profession, a vocation, and a gift to others and to the self. Holding on to these notions enables one to see through and beyond obstacles and difficulties.
(e) Above all, it is important to remember that one is not alone. By definition, medical care of a patient and teaching learners are collegial efforts of highly skilled persons who work toward a single end as a team. No one person can do everything alone. Cherish colleagues. Seek friends in team members. Do not let hasty words or arrogant appearances obscure the hardworking, earnest, kind soul inside the business-like exterior [8].

Love is a central theme in the profession of medicine and academics. It is an undeniable drive that leads individuals onward to serve others. Hallowed in the Oath of Hippocrates and in analogous promises, that love, that call on behalf of others, needs to be recognized and honored especially in periods of unfulfillment or discouragement. Naming that force for what it is and claiming it again as one's own turns the corner, away from discouragement toward renewal of the promise to that primal altruism, which first sparked the notion of becoming a physician, an academician, those first inchoate but undeniable longings to claim an honored tradition of service on behalf of and for others.

I will close with a story, a true story, that illustrates that dedication and the sometimes odd and sometimes strange way in which the oath, the promise, is mirrored back to us, so we can see it with renewed clarity and grasp it again with renewed firmness/determination. Each year a group of our fourth-year students goes to the Capital Medical University in Beijing to do a month long elective. One year in the not too distant past, a group of our students was talking informally with a group of Chinese medical students. At first, the conversation focused on medical practices, but then moved on to the differences and similarities in being a medical student in Beijing versus Houston. Our students began explaining the special terms used to describe noncompliant patients in the public hospitals. Terms such as "dirt-ball" and "scuz-bag" were named and described to the Chinese students. Definition was difficult since understanding these terms required a facility with slang most of the Chinese students did not have. Finally our students mentioned the term "G-o-m-e-r" for "get-out-of-my-emergency-room." The Chinese students brightened. "Oh yes," they said, "We have that word, gomer." Amidst the smiles of satisfaction on our students' faces for having broken down a communication barrier, our students asked, "What does gomer mean in Chinese?" "It means brother, the brother you *love* rather than your relative. It means brother," they said.

References

1. Viggiano, T.R., and Strobel, H.W. (2007) The Career Management Life Cycle: A Model for Supporting and Sustaining Individual and Institutional Vitality Proceedings of the Faculty Health Conference University of Texas Health Science Center, Houston, Texas.
2. Hubbard, G.T., and Atkins, S.S. (1995) The professor as person: the role of faculty well-being in faculty development Innovative Higher Education 20, 117–128.
3. Wang, D., Shaheen, J., Guze, P., Wilkerson, L., and Drossman, P.A. (2006) The impact of the changing health care environment on the health and well-being of faculty at four medical schools Academic Medicine 81, 27–34.
4. Nemko, M. Overrated career: Physician US News and World Report (Accessed January 2, 2008, at http://www.usnews.com/articles/business/best-careers/2007/12/19/overrated-career-physician.html).
5. Wall Street Journal, Vol CCL, No. 15. Page 1, columns 4, 5. July 19, 2007.
6. O'Connell, R.J. (2007) Time: research necessities make it hard to keep track Nature 450, Correspondence.
7. McKeown, R.D. (2007) Time: accounting problems caused by Caltech System Nature 450, Correspondence.
8. Litzelman, D.K., Williamson, P.R., Suchman, A.L., Bogdewic, S.P., et. al. (2008) Fostering faculty well-being through personal community and cultural formation at an academic medicine center: Indiana University School of Medicine as a case study (this volume).

Part V
Supports and Interventions

Chapter 13
A Model for Designing and Developing a Faculty Health Program: The M. D. Anderson Experience

Ellen R. Gritz, Janis Apted, Walter Baile, Kathleen Sazama, and Georgia Thomas

Abstract The University of Texas M. D. Anderson Cancer Center provides an approach to formulating a faculty health program that may guide others in organizational structure, in relation to executive management, and in programming. In 2000, following the suicide of a faculty member, concerned faculty formed a Faculty Health Committee (FHC). The Committee secured consultants to conduct focus groups, the results of which helped guide the Committee in producing educational programs for faculty and for assisting faculty leaders in recognizing stress, impairment, and burnout, and educating leaders on communication. Additionally, the Committee interfaced with other committees at M. D. Anderson relevant to their mission and arranged collaboration with Faculty Development to increase the range and possibility of programs. The Faculty Assistance Program (FAP), a key service that was established early and is still ongoing offers private and confidential psychological consultations off-site for faculty and their families at no cost. In 2005, the Committee selected a director to expand the programming, particularly by holding meetings to identify the stressors specific to each department and to provide seminars and other activities at departmental meetings on topics chosen as relevant. Inspirational and entertaining programs have also been

E.R. Gritz
Professor and Chair, Department of Behavioral Science, Olla S. Stribling Distinguished Chair for Cancer Research, The University of Texas M. D. Anderson Cancer Center (M. D. Anderson), Houston, Texas, USA
e-mail: egritz@mdanderson.org

J. Apted
Associate Vice President, Faculty Development, M. D. Anderson, Houston, Texas, USA

W. Baile
Professor, Department of Behavioral Science, Director, Program in Interpersonal Communication and Relationship Enhancement (I*CARE), M. D. Anderson, Houston, Texas, USA

K. Sazama
Professor, Department of Laboratory Medicine, M. D. Anderson, Houston, Texas, USA

G. Thomas
Associate Professor, Department of Infectious Diseases, Executive Director, Employee Health Services, M. D. Anderson, Houston, Texas, USA

T.R. Cole et al. (eds.) *Faculty Health in Academic Medicine*,
© Humana Press, a part of Springer Science + Business Media, LLC 2009

offered, such as social mixers, classical music, ethnic music, opera, and drama. These have been popular and effective ways of renewing and restoring energy and concentration. Program offerings are continually evaluated, in order to create the most effective structure with the broadest reach.

Keywords Faculty health, faculty well-being, academic health centers, stress, stress management, burnout, depression, work-life balance, resilience, self-care

Introduction

The Faculty Health Program at The University of Texas M. D. Anderson Cancer Center was designed from the outset to identify issues and provide programs for both clinicians and research scientists in the institution. While there is widespread acknowledgment of increasing stressors in the academic medical environment and the adverse toll these stressors can take upon physicians—specifically, in physical and mental health, patient care, job performance, productivity, and personal and family life—little or no attention has been directed to similar phenomena among research scientists working in these types of institutions. This chapter presents a model for the development of a faculty health program, based on the past seven years of effort at M. D. Anderson, authored by the key faculty and staff who initiated and built this program.

History of the Initiative for Faculty Health

In late 2000, the suicide of a renowned and highly successful surgeon at the University of Texas M. D. Anderson Cancer Center stimulated several faculty members to propose to develop a faculty health program that would address a broad range of issues related to faculty stress and dysfunction. Drs. Ellen R. Gritz, Georgia Thomas, and Walter Baile expressed to Dr. John Mendelsohn, President of M. D. Anderson, and his senior leadership team their strong concerns that a colleague could become this severely distressed without those around him either recognizing the signs or possibly not knowing what to do in response. Additionally, these faculty members were concerned that the distressed practitioner either did not acknowledge his problem, kept it to himself or, more likely, did not know where to turn for help. They also noted the absence of an institutional leadership acknowledgment of the nature of the death—an observation that was interpreted by some faculty as a denial of the significance of his suicide. Dr. Gritz and her colleagues proposed to form a group to address these issues.

With the strong support of the President and his executive management committee, the Faculty Health Committee was formed in January 2001. Dr. Gritz has chaired the committee since its inception, with the continuous participation of Drs. Thomas, Baile, and Sazama, among other representatives from major sectors of M. D. Anderson. Some members are permanent and ex-officio; others are in rotation. The Associate Vice President of Faculty Development (Janis Apted) and the Associate Vice President of Women Faculty Programs (WFP) (Elizabeth Travis, Ph.D.) serve permanently on the Faculty Health Committee and contribute to the planning, sponsorship, and implementation of creative, relevant programming.

As the next step in the development of the program, a working group chaired by Kathleen Sazama, M.D., J.D., then Vice President for Faculty Academic Affairs, drafted a comprehensive faculty health program, consisting of three separate but related concepts: (1) prevention of morbidity; (2) intervention for distressed faculty; and (3) response to catastrophic events such as faculty suicide. The responsibility for organizing prevention efforts was assigned to the new Faculty Health Committee, with the primary charge of educating faculty and the leadership. The responsibility for intervention fell primarily on the existing Employee Health department and on the medical leadership. At M. D. Anderson, Employee Health is a multidisciplinary department that includes an internal Employee Assistance Program (EAP). The EAP was instrumental in guiding the group as they decided how to address and manage impaired faculty. The third (and hopefully seldom invoked) concept, namely, response, was appropriately delegated to the President and his management committee. Their roles in responding to any new "sentinel event" are unquestioned. A protocol was developed to encompass all aspects of the organization's response to the unexpected death or significant injury to a faculty member, including key groups and topics, such as patients, colleagues, family, staff, security of the information, media response, continuity of activities, and the like. The experience with M. D. Anderson's faculty suicide showed that painful tasks, such as notifying current patients about his death, fell to peers who were already devastated by the event. One surgeon described poignantly how difficult it was to repeat the news of his friend's death over and over again to each affected patient. The response protocol is designed to reduce the amount of decision making needed in the wake of a tragedy like this. Many work areas within M. D. Anderson want to provide help and support after a faculty death. Thus, the response protocol will facilitate working together as effectively as possible, minimizing organizational turmoil.

In 2003, the Faculty Health Committee (FHC) was successful in obtaining approval for the addition of a full-time faculty member to serve as Director of Faculty Health, providing continuity for the committee's efforts and expanding the reach of the programs it offered. The FHC served as a search committee for a Director of Faculty Health, a position that would facilitate an increase in programming to respond to the 2005–2010 Strategic Vision for MDACC, especially Goal 7: to safeguard and enhance our resources—most importantly, our human resources. In November 2005, Thelma Jean Goodrich, Ph.D. joined the faculty of the Department of Behavioral Science as Associate Professor and Director of Faculty Health, to continue implementing the program in faculty health education, under the umbrella of the Faculty Health Committee.

Focus Groups

The formative stage of the committee's work consisted of the research, data-gathering, and information-sharing. The task was to understand the nature and scope of the problem. Aspects of programs at the University of Ottawa and Vanderbilt University provided ideas for M. D. Anderson. Apart from a few other academic health centers where stress management was occasionally offered to faculty, there were no well-developed, comprehensive faculty health programs.

After reviewing the literature and talking to colleagues nationwide, the Faculty Health Committee focused on learning about the specific stressors affecting M. D. Anderson personnel by listening to its own faculty. Pooling information, survey data, and focus group results about faculty and their work gave the committee a sense of the many causes and sources of stress throughout the institution and within the various clinical and research groups.

One of the richest sources of data was the focus groups sponsored in 2001 by the Chief of Psychiatry, Dr. Walter Baile. These group discussions focused on the personal aspects of being an oncologist–"the demands, the burdens, the rewards, the emotional costs" [1]. In four different and intense group discussions, one of which was held with surgeons two days after the suicide of their close colleague, it became clear that although treating people with cancer shares many similarities with clinical care in other disciplines, the problems are "intensified and augmented by our patients' feelings and fears about the diagnosis and by the social and personal attitudes and anxieties related to the word 'cancer' " [1]. In addition, the competitiveness of these highly trained physicians and their resistance to any feelings of self-doubt, sadness, or what they considered to be personal inadequacies motivated them to keep a very tight rein on their emotions and reactions.

Transcripts of these focus groups became the script of a video and workbook titled "On Being An Oncologist" which featured William Hurt, the well-known actor, and Megan Cole, the actress who originated the role of Vivian in "Wit." (Wit is an evocative play about a scholar of 17th-century English literature who experiences the transformation of her controlling nature as she succumbs to cancer and faces death.) Using the words of our M. D. Anderson's clinicians made the video a powerful tool for instruction about the stressors of oncology. The video and workbook set were distributed by request to over 8,000 medical schools, academic health centers, and practicing physicians, worldwide. The overwhelming feedback confirmed that high levels of stress were endemic to the field of oncology, and perhaps to medicine in general.

For physicians, the constant anguish of dealing with dying patients and the feelings of grief that go unexpressed is a powerful dynamic. The sense of urgency, high expectations, and the desperation of patients and their families carried over to clinicians with already internalized and often perfectionistic standards. Two recent surveys of cancer specialists reported significant rates of burnout in the oncology community. The first, a study of 549 surgical oncologists revealed a rate of 28% [2], while a study of 1,740 clinicians (mostly hematologists and medical oncologists) showed a rate of 61.7% [3]. Burnout is associated with decreased job performance, including suboptimal patient care, reduced job commitment, stress-related health problems, and low career satisfaction. Several studies have reported positive outcomes from intervention programs to decrease burnout among physicians [4, 5]—see also Chapter 3.

To probe further into faculty health issues, the Faculty Health Committee contracted in early 2003 with two nationally known consultants whose expertise was in physician stress and wellness. Their discussions with faculty identified the following concerns:

1. Concern about the relationship between reaching the institution's budgetary margin and work overload.
2. The growing rage being felt by faculty who experienced stress as an uncontrollable "train wreck."
3. Constant multitasking in a matrixed organization, such that faculty frequently had reporting responsibility to more than one leader.
4. Rapid institutional growth that was perceived as chaotic.
5. The sense of never having enough time.
6. Idiosyncratic institutional infrastructure systems and the frustration of navigating what appeared to be Byzantine policies and compliance issues.
7. Fear of inadequacy regarding selected professional assignments or tasks and the perceived need to hide that feeling.
8. Inability to discuss these issues with chairs or colleagues for fear of appearing weak.
9. The request by leadership to increase productivity by 15%. How would faculty accomplish this?

A special focus group with the Faculty Spouse Association was especially interesting and poignant. Among their concerns and insights were these thoughts:

1. Because of patient confidentiality, spouses hear little about the patients and therefore do not always know what their partners are dealing with.
2. M. D. Anderson was a powerful environment with exciting work going on. Sometimes, being at work appeared to be more interesting than being at home with the spouse and kids.
3. Guilt about asking partners to attend their children's school functions or to take vacations.
4. The inability to understand the nature of the stress because they did not feel welcome at M. D. Anderson, even as volunteers.
5. Partners countering demands to spend more time at home with "I'm looking after very sick cancer patients," an effective and guilt-inducing way of shutting down dialogue.
6. The need to accept a workaholic partner or walk away.
7. Concerns that when partners retired, they would have no other life, or would infringe on ongoing activities of the spouse.

Later in 2003, another expert on stress in the work lives of clinicians and research scientists was brought in to conduct educational sessions for the chairs and other faculty leaders and to help the committee gather more data. Dr. Edward Nace, was chosen for his breadth of experience in physician health, including his tenure as a past chair of the Texas Medical Association's Physician Health & Rehabilitation Committee. His focus groups identified a number of specific stressors for faculty including: the burdens of heavy patient care schedules and emotional strain associated with caring for very ill patients; the pressure and uncertainty of grant funding; poor communication between top management and faculty; the feeling of powerlessness in a large and growing institution; the frustration of dealing with organizational politics; time pressures; and the lack of amenities for faculty such as dedicated gyms and dining facilities. Focus groups have identified other critical issues and continue to be used as a means of tracking and identifying the most prevalent sources of stress.

Dr. Nace concluded that the stress levels of faculty were high and reflected the culture of "triple threat" lauded in academic circles: the ideal of being a productive researcher, an astute clinician, and an able educator. Faculty are internally driven and highly motivated, he pointed out, and like many in academia they tend to blame the administration for their stress. However, his opinion was that much of the stress was self-induced and would occur even under a "very benign administration." Perfectionism, compulsiveness, and a tendency to be dismissive and secretive about personal vulnerabilities were all at play in this highly competitive environment. The challenge was how to help faculty become less isolated and more willing to seek and accept support.

We conducted another set of focus groups to explore whether faculty from other cultures experienced unique stressors in our institutional environment, and how they coped with them. The surprising outcome was that these faculty were stressed by the same things as their American-born counterparts: namely pressing time demands; high patient loads; constant grant writing with uncertain outcomes; and frustrating layers of bureaucracy. M. D. Anderson was perceived as very multicultural and supportive. As we move forward with programming, focus groups conducted by Faculty Development and the Women Faculty Program may identify other faculty stressors, as well.

An all-employee survey was conducted in 2005. The results from faculty completing that survey plus findings from focus groups, department meetings, and other qualitative sources of data will guide the formulation of a faculty climate survey in the near future (see Section on Future Plans and Conclusions).

Educating Faculty and Faculty Leaders

Dr. Nace returned to M. D. Anderson about six months after his first visit to give several talks on stress, burnout, and physician impairment to chairs of departments and other institutional leaders, who participated in large numbers. Not only were they concerned for their faculty, but they themselves tend to be the epitome of stressed individuals—tired, overloaded, dealing with growing numbers of faculty and staff, frequently feeling powerless, and faced with constant decision making while trying to balance their own academic careers with demanding administrative responsibilities. Not surprisingly, Dr. Nace uncovered a strong need among these faculty to talk about and discuss their own stressors. The department chairs were energized by their contact with him, leading to long discussions after his formal lectures had ended. Ongoing programs for faculty leaders are critically important to the overall success of the Faculty Health Program. Not only do these leaders need to understand more about their own stress so they can develop better coping skills, they also need to be better prepared to identify faculty who may need assistance and encourage them to seek help.

A program that has received high marks by faculty is the Faculty Leadership Academy (FLA) and its related programs, run out of the Faculty Development office. Designed to develop the critical skills needed to manage and lead people, the program is now in its sixth year and has been successful in developing more confident, capable faculty leaders. Because of its popularity, enrollment has become competitive, and condensed versions of the FLA are being offered to junior faculty who have yet to assume formal leadership roles.

Small groups of faculty who are willing to talk about their personal lives have enriched our Faculty Health curriculum. The Faculty Health Program sponsored two such panel discussions, one on the experience of faculty who have dealt with personal illness and the second on faculty marriages. Both were well-attended. Our faculty panelists volunteered to participate, providing touching and genuine snapshots of their lives. The presence of executive leadership at one of these panel discussions demonstrated a commitment to faculty health in vivid and personal terms. M. D. Anderson's president, Dr. Mendelsohn, spontaneously decided to attend the panel discussion on chronic illnesses. After the panelists had concluded, he raised his hand and spoke movingly of his support for these faculty members, and of his personal confidence in them as they continued in their careers. By his presence, he illustrated M. D. Anderson's commitment to faculty health in an unrehearsed but powerful manner.

Overview of Programming

The vision of the Faculty Health Program is to enhance the well-being of our faculty by helping them to achieve professional and personal satisfaction during their careers at M. D. Anderson. Work-life balance, physical and emotional health, and peer support are core components of faculty health.

To achieve this goal, we have evolved a multifaceted program of services to individuals, to departments, to academic leaders, and to the institution as a whole. The plan for formulating such a program includes three phases, the first two of which have been discussed in the foregoing paragraphs: (1) obtaining benchmark data, acquired through use of focus groups, surveys, and interviews with key stake holders; (2) applying well-researched principles of stress management, and adopting similar programs from corporations and the limited number of other medical institutions that address faculty health; and (3) holding meetings between the Director of Faculty Health, Dr. Goodrich, and Department Chairs and faculty, often scheduled in connection with regularly occurring departmental meetings. The purpose of these meetings is to collect current information on the major sources of perceived stress as well as the sources of strength and support that exist in each department.

Visits to department chairs and division heads by the Director of Faculty Health is a critical way to make the issues of faculty health and the range of institutional responses known to faculty leaders. Because the institution is so large with nine divisions and 56 departments, these visits occur over a two-year cycle. For the Director, these conversations are rich opportunities to gather information on how faculty health, stress, and resilience are viewed by the leader and whether any measures have been taken to address faculty well-being. There are, in fact, some enlightened chairs and division heads, but there are also those who either deny the problem or take the old school approach of "this is academic medicine, this is what I had to face when I came into the field, I hire only the best who know what they are getting into."

In many cases, the Director of Faculty Health is asked to return to do presentations at faculty meetings. The most frequently requested topics are how to set priorities and keep them, quick techniques for managing stress throughout the day, and how to successfully integrate work and personal life. The challenge of managing work and having a home life is consistently raised by early career faculty as the issue of greatest concern to them. Other topics include handling the emotional strain of caring for the critically ill, sleep disturbances, time management, learning to say "no," communication issues, organizing skills, and professional renewal. Conversations on these topics allow faculty and their leaders to identify the burning issues. Are faculty feeling overwhelmed? Do they feel they are managing their stress well? Do they want more support and help with workplace demands? Do they feel safe in talking openly about these things in front of their chair? Responses in the different departments vary widely. Written reports on each of these sessions are shared with the Faculty Health Committee leaders who use them as sources of data for planning purposes.

To improve understanding and increase awareness of the importance of a healthy faculty, the Faculty Health Committee has sponsored over 60 programs to date, many of them in collaboration with Faculty Development. The Faculty Health programs are by necessity diverse and include some designed for large audiences along with many "boutique" programs for specific, small faculty audiences (see Table 13.1 for a selected list of programs from 2001 to present). A "one size fits all" program would not be effective in such a large and rapidly growing institution. A continuing assessment of needs ensures that programming is on target, and evaluation of all sessions provides feedback on whether program objectives were met. In addition, evaluations allow faculty to tell us what other programs they are interested in attending.

Some highlights of programs are given here, and others appear in subsequent sections below. As an example of one type of educational program, some faculty

are seeking information on stress management and burnout, or want to develop skills to deal with stress. Thus, we have engaged prominent speakers (such as Glen Gabbard, M.D., a psychiatrist who is a nationally recognized expert in faculty stress issues) to address the topic of "The Perils of Perfectionism: Stress, Burnout and Depression"[6]. Stress management and stress reduction techniques are taught in a variety of formats.

A more "practice-oriented" program is the weekly Sudarshan Kriya session. On-site workshops for this yoga breathing technique, introduced to faculty in 2005, resulted in sustained interest and regular sessions. Weekly meditation and Tai Chi groups are led by expert practitioners and teachers. Other groups of faculty have

Table 13.1 Selected faculty health programs by major topic area (2001–2008)

Communication skills
The Human Touch: Skills for Thriving in the Technological Age
Complicated People: Keys to Coping with Counterintuitive Behavior
Responding to Emotions: Clinical Communication Skills Workshop Series
The Dance of Connection: How to Talk to Someone When You're Upset

Work flow and productivity
What's Your Organizing Problem?
Getting Things Done: The Art of Stress-Free Productivity
Overloaded Circuits: Why Smart People Underperform

Stress/resilience and work/life balance sessions
Faculty Quality of Life: An Endangered Topic
Keeping Sane in an Insane World: Stress Management for Stressed Faculty and Chairs
The Perils of Perfectionism: Stress, Burnout, and Depression in Professionals
Balancing Career and Relationships: An M. D. Anderson Faculty Panel
Resilient Scientists, Physicians, and Academic Medical Families: Keeping the Flame Alive
Maintaining Faculty Satisfaction in the Face of a Broken Health Care System: Why You Need to
    Invest in Personal Relationships to Succeed in Academic Medicine
The New Medical Workplace: A Session for Department Chairs
Field Notes on the Compassionate Life

Meditation and yoga
Introduction to Kriya Breathing
Yoga Breathing: Neurophysiology and Health Benefits for Stress, Anxiety, and Depression
Happiness: The Guide to Developing Life's Most Important Skill
Mental Fitness and Peak Performance at Work
Experiential weekly sessions: Kriya Breathing, Meditation, Tai Chi

The arts and humanities: cultural events and self-reflection
The Skills of Empathy: A Series of Talks Incorporating Poetry and Literature
Balancing Engagement and Detachment: Skills from the Dramatic Arts
Wit: A One Hour Version of the Play
Music and Illness: Annual Piano Concert and Lecture
Renewing the Spirit in Work
Trio Colombiano de Cuerdas: Chamber Music Performance
Life and Death in Opera: Great Musical Moments. Houston Grand Opera
Art and Healing: A Tour of the Menil Collection
Classical Indian Dance Recital: Anjali Center for Performing Arts

learned mental fitness and peak performance meditation and relaxation skills from two expert teachers from Seattle who continue to bring their unique form of teaching to M. D. Anderson's faculty and staff. As more and more faculty participate in these types of programs, follow-up programming is offered so that a community of faculty practitioners is created who can continue to learn together.

Cultural events are very popular with faculty, allowing them an enjoyable hour away from their work. Small discussion groups on literature and medicine have facilitated interaction and communication among colleagues, often leading to sensitive interchanges. "Stress-busting" entertainment has drawn large audiences. Two of our most popular events have been musical. Dr. Richard Kogan, a New York-based psychiatrist who is also a concert pianist, gives an annual piano concert and lecture on the music and life (including medical and psychiatric illnesses) of a famous composer. Performances have focused on the music and life of Gershwin, Tchaikovsky, Beethoven, Bernstein, and Mozart. Second, young artists from the Houston Grand Opera (HGO) Studio performed a series of arias and duets from 19th and 20th century operas, portraying illness and death. "Life and Death in Opera: Great Musical Moments" featured selections from La Traviata ("consumption," i.e., tuberculosis), La Bohème (tuberculosis), Death in Venice (cholera), and Angels in America (AIDS). Historical and cultural interpretations were provided by HGO Dramaturg Colin Ure and HGO-co (outreach) Director Sandra Bernhard. The success of this event has led to a decision to feature it as an annual event, as well. Performances from ethnic traditions have included music by a Colombian trio, and a Bharat Natyam dance exhibition given by the world-famous Indian classical dancer, Ms. Rathna Kumar.

Web Site

A Faculty Health web site serves as an educational resource, a source of information, and a place for faculty to assess their health and well-being, stress levels, depression, and burnout. Future programs are noted for planning purposes. There is also a list of wellness resources available throughout the Texas Medical Center such as Weight Watchers, the Institute for Religion and Health, the University of Texas Fitness Centers, and arrangements with other local fitness centers where University of Texas faculty and staff can receive membership discounts. The web site also offers help to faculty in setting up special interest groups for recreational activities such as scuba diving, dancing, book clubs, and sailing.

Faculty Assistance Program

As described above, the Faculty Health Committee has arranged presentations for institutional leaders, including division heads and department chairs, to aid them in identifying faculty in distress and to suggest ways to encourage those faculty members to seek help. Further, the Committee supported the establishment of a mechanism to maintain institutional anonymity for faculty who receive help through the various assistance programs. Although faculty had been using the internal Employee Assistance Program (EAP), faculty utilization was only 1.5%, compared to non-faculty utilization of 5% per year. To address this need, we established a Faculty Assistance Program (FAP) to provide individual consultation outside of M. D. Anderson, using community mental health practitioners. Faculty can choose from among four FAP consultants, all of whom are doctoral-level therapists. A faculty member and/or immediate family member calls the therapist directly, and the institution is billed without revealing the faculty or faculty family member's identity. As part of their orientation, all new faculty are informed of this service. For faculty in our off-site facilities near Austin, two psychologists in that community provide Faculty Assistance Program services. The FAP has both an Intranet and an Internet web page, eliminating the need to contact anyone inside M. D. Anderson to access services.

The psychologists who participate in the FAP were chosen specifically for their reputations and for their interest in the issues of high-performing professionals. All of them were personally known to the Directors of the EAP and of Employee Health (Dr. Thomas). Although Faculty Health receives no information on specific clients, utilization statistics make it clear that this adjunct service has reached additional faculty and their family members (Fig. 13.1).The total number of faculty served is still modest yet, together with the EAP, we have more than doubled the number of faculty who have used these two programs. The gender distribution and proportion of Ph.D. /M.D. participants is representative of our faculty population as a whole. Although the program is open to family members as well, they constitute a minority of these self-referrals (data not shown). Faculty Health committee members, including Dr. Goodrich, meet at intervals with these consultants to understand how

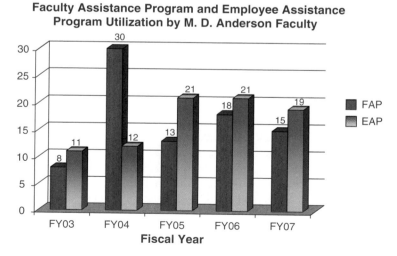

Fig. 13.1 Faculty assistance and employee assistance program utilization by M. D. Anderson faculty

we can better help our faculty. Job stress, conflict within family systems, and psychological disturbances, particularly depression, have been mentioned most often as problems seen by the FAP consultants. The FAP is an assessment and referral model, providing only three to four free sessions. Many faculty elect to continue counseling, either with these providers or with a referral that the FAP consultant has facilitated. Some faculty have entered group as well as individual therapy as a result of using the FAP. One of these therapists has a men's group that recently went camping in Big Bend National Park. Even though experiences like this one may touch only a single faculty member, we have seen faculty who share their sense of renewal with their peers, helping to spread a message of optimism and hope.

Although some faculty clearly find the FAP a useful service, anecdotal comments by physicians in a number of different departments highlight a recurring theme, namely, that some of our faculty are reluctant to access any mental health assistance that is connected to M. D. Anderson. Reasons given range from the belief that the service is not truly confidential, to the possibility that a mental health diagnosis may be disclosed to the Texas Medical Board (TMB). Currently, the TMB will open an official investigation on every licensee who reports or is found to have a major mental illness. Such investigations can result in a public Board Order. The threat of an event like this is a powerful disincentive toward getting help.

A specialized type of support is offered by an Alcoholics Anonymous group that meets twice weekly in our main ambulatory clinic building. This group includes some faculty in recovery who reach out to their peers. As programs at other institutions have commented, faculty who are in recovery from substance use disorders can be a valuable resource for faculty health.

Other Programs for Faculty

Additional programs sponsored by a comprehensive Faculty Development Program dovetail with the Faculty Health programs by helping faculty develop the vital skills they need to feel competent in their work, including the ability to confidently handle difficult physician–patient interactions, the communication skills needed to work effectively with diverse faculty and staff and to resolve conflict, team performance skills, and management and leadership knowledge, and expertise. Because so much stress accrues from feeling inadequately equipped to deal with the challenges of complex careers in science and medicine, the Faculty Development programs are targeted to specific stages of faculty careers, from the earliest experience offered through the Junior Faculty Development Program to the most advanced levels of leadership development that is the focus of the Faculty Leadership Academy (FLA) and its related programs. These unique leadership development courses have been designed in partnership with M. D. Anderson faculty by a leading management development consulting group.

Faculty Development also provides crucial and varied educational programs and resources focusing on interpersonal skills (Interpersonal Communication and

Relationship Enhancement known as I*CARE), a Classrooms of the 21st-century teaching skills program, a Faculty Mentoring Advisory Center, and one-on-one coaching sessions for faculty and their staff who wish to conquer disorganized work spaces, files, desktops, and PDAs. All these initiatives have been developed from ongoing needs assessments of faculty, identifying problems and areas of discomfort that faculty feel when faced with new challenges in their work.

Another facet of the work of Faculty Development is working with intact faculty teams in clinical and research departments to enhance team performance and reduce or eliminate team dysfunction. Team dynamics contribute enormously to each individual faculty member's sense of well-being. Dysfunctional teams can cause an infection of bad behavior, poor attitudes, and negativism, all of which contribute to increased faculty stress and lowered productivity. Co-facilitating team alignments with the Faculty Leadership Academy's leadership development consultants, Faculty Development staff is active in addressing team performance issues across the institution.

The Women Faculty Program (WFP), directed by Elizabeth Travis, Ph.D. (a faculty member who is an Associate Vice President) functions in cooperation with Faculty Development and Faculty Health to address the specific issues of women in academic medicine and to foster the development of talented women academics. The WFP provides career development, skill-building, and networking programs for women faculty and educational opportunities for faculty chairs to enhance their understanding of the particular challenges facing women and to allow them to address the needs of women faculty more effectively.

The relatively new Ombuds Office is another service well utilized by faculty. Anu Rao, Ph.D., the director of the office, is a member of the Faculty Health Committee and provides valuable insights on the difficulties faculty experience in the workplace. A confidential resource, the staff of the Ombuds Office provide neutral perspectives on incivility, harassment, discrimination, and authorship, and work toward achieving win–win solutions.

In addition to the resources described above, the Faculty Health program is supported by an Employee Assistance Program (EAP), which has a strong reputation as a consultative tool for supervisors. Faculty department chairs can access the EAP directly for help with a troubled faculty member. More commonly, a faculty member will be referred to the EAP by a senior faculty administrator or by a member of the Faculty Health Committee. The EAP notes that these supervisory referrals include a mix of acute impairment as well as other behaviors indicative of faculty distress. As in nondoctoral adult populations, there is a significant amount of untreated mental illness, particularly depression.

Finally, a Practitioner Peer Assistance Committee (PPAC), chaired by Dr. Baile, has been recently established to help address the needs of all licensed practitioners, both faculty and non-faculty (with the exclusion of nurses, who have their own infrastructure for such situations). This committee's mission is to intervene with distressed practitioners. It provides an important connection between prevention and intervention, as well as meeting the guidelines of The Joint Commission's Medical Staff standard.

Figure 13.2 depicts the relationship of Faculty Health to other M. D. Anderson programs.

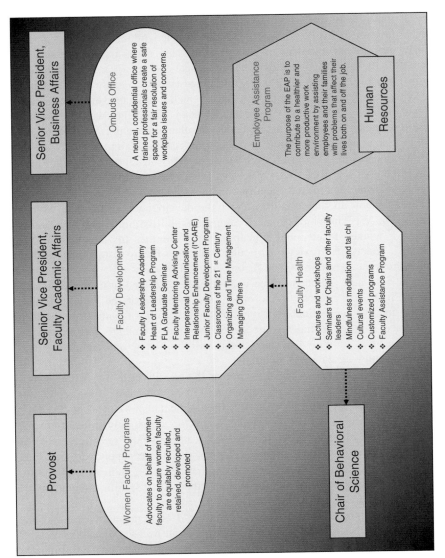

Fig. 13.2 Relationship of faculty health to other M. D. Anderson programs

Future Plans and Conclusion

At this point, the Faculty Health Program offers diverse services and programming and has strong relationships with related institutional programs serving faculty, as summarized above. The next step is to conduct an in-depth survey of the faculty, similar to that undertaken at Vanderbilt University [7] and at the University of Ottawa School of Medicine (see also Chapter 4). This survey, planned for FY 09 (September 08–August 09) will provide current data on the perceived health and functioning of the faculty, estimates of impairment and burnout, suggestions for filling gaps in programming, and ideas for new initiatives. These data will supplement the information gathered in department-specific meetings and workshops, program evaluations from ongoing presentations, themes mentioned by Faculty Assistance Program practitioners, and issues arising in the literature.

A new round of educational sessions is already planned for faculty leaders, in particular, department chairs, section heads, and other faculty administrative leaders. A strong influx of new leaders and recruits to M. D. Anderson makes this imperative. The skills needed to identify and communicate with faculty at risk or suffering impairment need to be practiced and, as was learned from Drs. Nace's and Gabbard's original presentations, are not part of the routine education of faculty leaders. In addition, faculty leaders should be encouraging communication about stressors in order to prevent the development of high risk and impairment. This is also a skill that requires education and practice, since it is far easier to avoid raising or discussing such topics.

The Faculty Health Committee has become aware that faculty stress and wellness issues are often directly related to institutional policies, practices, and organizational culture. Innovative programming needs are already identified on a spontaneous and/or reactive basis. In addition, as described above, collaborations are proactively developed with other institutional committees and offices that have important roles with faculty, including the Practitioner Peer Assistance Committee, the Ombuds Program (see also Chapter 15), and the Women Faculty Program. The Committee has a representative from each of these areas, as well as representation from, and strong relationships with, additional important institutional resources, such as Faculty Academic Affairs, the Faculty Senate, Human Resources, and the Office of Institutional Diversity. In this way, the Faculty Health Program seeks a dynamic relationship with our faculty and its leadership arms.

For the good of the faculty, for the good of the institution, and for the good of the patient population we care for, a strong faculty health program is essential. Some sources of stress, once identified, can be changed, removed, or mitigated. In situations where sources of stress cannot be changed, programs are offered that encourage healthy responses to stressors and teach restorative practices. Evaluation of programs for their appeal and their effectiveness provides guidance so to functioning in an evidence-based manner and, in addition, sharing findings with the wider professional audience.

References

1. Baile, W.F., and Buckman, R. (2002) On Being an Oncologist: Reflections on the Personal Dimensions of Clinical Oncology (video and workbook) Houston, TX: The University of Texas M. D. Anderson Cancer Center.
2. Kuerer, H.M., Eberlein, T.J., Pollock, R.E., et al. (2007) Career satisfaction, practice patterns and burnout among surgical oncologists; report on the quality of life of members of the surgical oncology society *Annals of Surgical Oncology* **14**, 3029–3032.
3. Allegra, C., Hall, R., and Yothers, G. (2005) Prevalence of burnout in the US oncology community: Results of a 2003 survey *Journal of Oncology Practice* **1**, 140–141.
4. Dunn, P.M., Arnetz, B.B., Christensen, J.F., and Homer, L. (2007) Meeting the imperative to improve physician well-being: assessment of an innovative program *Journal of General Internal Medicine* **22**, 1544–1552.
5. Le Blanc, P.M., Hox, J.J., Schaufeli, W.B., Taris, T.W., and Peeters, M.C.W. (2007) Take Care! The evaluation of a team-based burnout intervention program for oncology care providers *Journal of Applied Psychology* **92**, 213–227.
6. Gabbard, G.O. (1985) The role of compulsiveness in the normal physician *Journal of the American Medical Association* **254**, 2926–2929.
7. Yarbrough, M.I., Ragan, P.W., Kendall, J.W., and Byrne, D.W. (2006) The faculty and physician wellness program work/life connections—EAP (presentation) International Association of Employee Assistance Professionals in Education Annual Conference, Nashville, TN, November 17, 2006.

Chapter 14
Fostering Faculty Well-Being Through Personal, Community, and Cultural Formation at an Academic Medical Center: Indiana University School of Medicine as a Case Study

Debra K. Litzelman, Penelope R. Williamson, Anthony L. Suchman, Stephen P. Bogdewic, Ann H. Cottingham, Richard M. Frankel, David L. Mossbarger, and Thomas S. Inui

Abstract Indiana University School of Medicine is the site of a unique initiative impacting faculty well-being through a comprehensive cultural change effort. The initiative was based on applying relationship-centered care not only to doctors and patients, but also to all members of the academic community across missions.

D.K. Litzelman
Associate Dean for Medical Education and Curricular Affairs, Indiana University School of Medicine, Indianapolis, Indiana, USA
e-mail: dklitzel@iupui.edu

P.R. Williamson
Department of Medicine, The Johns Hopkins University School of Medicine, Relationship Centered Health Care, Baltimore, Maryland, USA

A.L. Suchman
Relationship Centered Health Care, Department of Medicine, Department of Psychiatry, University of Rochester School of Medicine and Dentistry, Rochester, New York, USA

S.P. Bogdewic
Department of Family Medicine, Office of Faculty Affairs and Professional Development, Indiana University School of Medicine, Indianapolis, Indiana, USA

A.H. Cottingham
Director of Special Programs, Indiana University School of Medicine, Indianapolis, Indiana, USA

R.M. Frankel
Department of Medicine, Indiana University School of Medicine, Richard L. Roudebush VAMC, Regenstrief Institute Inc., Indianapolis, Indiana, USA

D.L. Mossbarger
Regenstrief Institute, Inc., Indiana University School of Medicine, Indianapolis, Indiana, USA

T.S. Inui
Associate Dean for Health Care Research, Professor of Medicine; President and CEO, Regenstrief Institute Inc., Indianapolis, Indiana, USA

T.R. Cole et al. (eds.) *Faculty Health in Academic Medicine*,
© Humana Press, a part of Springer Science + Business Media, LLC 2009

Early cultural change efforts focused on changing the formal curriculum and creating a broadly distributed, written document on the organization's guiding professional values. These initial cultural change efforts were foundational for a more comprehensive movement emphasizing formation as described in this chapter. Over several years, a wide variety of programs and professional development opportunities emphasizing personal formation (knowing self), community formation (finding community), and cultural formation (creating value) were offered. The insights, perceived impact, and observed outcomes are reported through the personal stories of Indiana University School of Medicine (IUSM) community members.

Keywords Faculty well-being, cultural change, academic health centers, informal curriculum, relationship-centered care, personal formation, community formation, cultural formation, faculty stories

Lost Bridge

Stanza Eight

No one remembers now why they called it Lost Bridge. No map
goes back that far. At dawn one October morning, a year ago,
I rowed out through the mists to a point half a mile beyond the bluff
And watched a hundred sandhill cranes lift from the water
and circle around the island, keeping the trees between us until
I finally came about. Eventually they settled down again
in the same place they've been stopping for thousands of years
on their migrations back and forth between the Gulf of Mexico
and Hudson Bay. That particular place, once a river, now a lake
that waxes and wanes, is still a fulcrum for their journeying,
a balancing point. I began to row back toward the landing, facing
the ridge where the sun would rise. The lake was so calm
the early morning light spread out over the mist-whitened surface
in one vast, unbroken gleam. For a moment, beneath the boat,
the water turned clear, to the farthest depth, and I could see
how I was moving along the original course of the river,
following and yet still not being determined by those twists
and turns, those long-forgotten shapes and contours.
Nor was there a current carrying me now. Suspended, borne up
by some invisible stillness, I moved under my own power.
The bluff at Lost Bridge came into view, half of its trees
and weedy coves still dark with shadow. Slowly, oarlocks
creaking, each dip of the oars trailing swirls in the water,
I left that world behind, and rowed across the shining lake.

(Jared Carter [1])

Introduction

Academic medical centers are a "fulcrum" for health journeys. Patients come to these centers of excellence with complex illnesses in hopes of benefitting from medical care beyond what might be expected in ordinary community settings. Future health professionals come for education and training that launches them into their careers. Faculty flock to academic medical centers because they love to teach, do research, or care for patients with challenging conditions.

Institutions with their own history, cultural norms, and patterns of activity, academic health centers may today seem "under water", burdened by demands for productivity, high throughput, high quality, and the call for ever-increasing revenues. The administration, faculty, students, and patients may be aware of the historic importance and core missions of these institutions, but these missions seem lost in time, visible only at moments, shrouded by mists, and perceived only as "weedy coves still dark with shadows". Some in these organizational communities may have no clear understanding of the place in which they have cast their professional lives and how all the members of this community can respond to the values that called them into medicine in the first place. Rowers across the shining lake, how shall we get our bearings?

Darrell Kirsch, in his 2007 Presidential Address to the Association of American Medical Colleges [2], acknowledged faculty members' "deep disillusionment regarding our ability to advance our core missions" within academic health centers. He postulated that the root source of this discontent is an imbalance between the energy and focus placed on strategic planning at the expense of attention to the culture of our organizations. Kirsch defined culture as "the shared values, assumptions, norms, behaviors, and rituals developed by a group, as well as the structures used to preserve these essentials". Just as conscious attention to our organizational cultures has helped to elucidate the hidden curriculum of medical education [3], so the examination of our academic cultures can provide insight into the work-world of our academic faculty—including what is impacting their job satisfaction, work performance, and overall perceptions of well-being. In his address, Kirsch suggested that the current dominant cultures of our academic health centers must change. Academic excellence, historically predicated on autonomous pursuits and independent achievements, now requires collaboration, team performance, and shared reward systems. Kirsch refers to Parker Palmer's book, The Courage to Teach [4], which addresses the harmful and unhealthful impact of living a "divided life" where one's personal values and professional behavior are divided by different expectations. An ideal academic culture would foster alignment between personal and institutional values including trust, collaboration, and shared accountability, so what we practice and how we behave embodies our shared values.

Is this kind of fundamental change—a shift in the culture of our institutions—likely or even possible? It may not be as far-fetched as one might first suppose.

Academic health centers have and will continue to attract individuals who are deeply caring and compassionate health care providers, highly motivated and innately curious intellectuals capable of incredible new discoveries in the business of educating open-minded and open-hearted trainees eager to grow into roles where they can serve others. The communities that evolve in these centers are not essentially static or predetermined, but are importantly influenced by the unique contributions of each member. The culture that prevails in these communities consists of patterns of relational behavior and attitudes that arise, propagate, and evolve within the medium of ongoing human interaction [5]. The complementary and synergistic effects of each member's strengths, weakness, style, and life perspectives collectively affect and define the culture of the institution.

How do the thoughts and actions of one affect the many? At any one point in time, the cultural patterns in an institution may seem, and be assumed to be, immutable. Yet even one questioning or uneasy individual, especially one such person among a group of intelligent and committed people, can disturb a standing pattern of interaction and generate new possibilities for reshaping the everyday work-world. Discomfort, and even pain, that arise when incompatibilities exist between a person's deepest values and sense of self and perception of the organization's values and purpose can be a spark that triggers questioning, reflection, and eventual change. The first courageous individuals who speak out often find they are not alone in their concerns, but are joined by others whose energy and commitment lead to a collective sharing of ideas that lead to change. Deep cultural change, also referred to as *cultural formation*, occurs when its members transform an organization, so that its purpose and values reflect the purpose and values of its people. Like-minded and like-valued people, we suggest in this chapter, can come together to create a nurturing, collaborative, energizing, and even healing culture that supports, rather than harms, its members.

As Palmer suggests, cultural formation or re-formation rests upon processes of *personal formation* that enable members of a community to reconnect with their personal values, beliefs, and strengths, and improve their abilities to relate to others in positive, life-enhancing ways, that is, living divided no more. Carefully constructed opportunities for faculty to engage in self-reflection can lead to clarity about personal strengths, weaknesses, values, and purpose. Practices such as deep listening and supporting colleagues through asking open, honest questions, rather than offering a "quick fix," can also be developed with time and practice.

Over a decade of work focused on changing the culture, including the adoption of a competency-based curriculum and the establishment and broad distribution of a statement on Indiana University School of Medicine's (IUSM) Core Values and Guiding Principles at IUSM, are described elsewhere [6,7]. These initial cultural change efforts were foundational for a more comprehensive movement emphasizing formation.

In this chapter, we provide an introduction to the concepts of *personal, community, and cultural formation* as foundational cornerstones for the cultural re-formation efforts that have taken place at IUSM. We report how a small cadre of individuals, who felt the discomfort of being part of an organizational culture that was perceived by patients, trainees, and staff as uncaring, uncommitted, selfishly competitive, mercenary at worst or unaware and disaffected at best, set about changing the culture of their large, academic medical center. The various programs and activities offered to IUSM faculty as part of the Relationship-centered Care Initiative (RCCI) and other, allied faculty development efforts intended to positively influence the professional culture of our organization are briefly described. An analysis of the stories solicited from faculty engaged in the various organizational change programs and activities provides a window into our academic community members' personal insights, the perceived impact of the organizational change process, and the outcomes observed by faculty participants. We close with lessons learned from our IUSM case study that might be applicable to other institutions and may have significance elsewhere in a general effort to assure faculty well-being and the vitality of our institutions.

Introduction to Personal, Community, and Cultural Formation

A large body of literature exists on the theories supporting personal, community, and cultural formation including numerous publications by Parker Palmer [4,8], the work of Ralph Stacey and his colleagues on Complex Processes of Relating [5,9–11], work in Appreciative Inquiry (AI) [12,13], and practical applications of these theories in health care, medical education, and administrative environments [6,14–16], which cannot be adequately condensed in this chapter, but provide excellent references for those planning to engage in organizational change work. A very brief introduction to these concepts and theories is provided below.

Personal Formation

Central to faculty health and well-being is knowing and caring for oneself. Being self-aware requires an active, mindful process of ongoing reflection. Deep reflection is facilitated by creating tranquil space and time (even short periods) set aside to check and, if necessary, recalibrate one's inner compass. The place and time for reflection can be facilitated through creative outlets such as journaling, painting, poetry and/or narrative readings, and conversations. This process allows one to stay in touch with personal strengths and weakness and the paradoxes of life events that promote joy and sorrow in the search for what is truly energizing and renewing. Although this is a very personal process, as social beings our understanding of self is made more fully whole through relationship with others and therefore can be practiced both alone and with others who can help us see ourselves more completely. The process and practices of knowing self have been termed "personal formation" with traditional roots in spiritual development, but with far broader application to professional and leadership development [4, 8]. Many relationship-centered care activities at IUSM have been designed with space and time built in for personal formation.

Community Formation

When like-minded individuals in organizations come together to help and support one another in creating shared values within the workplace they are engaging in community formation. Those who previously minimized the importance of their roles within the organization can emerge as "new-style" leaders, encouraged by the knowledge that they are not alone and/or by the welcoming acceptance and encouragement of community members. Some members come to the realization that their work units are not promoting healthy and professionally satisfying work environments and in these situations, the larger community can help the individual muster the "courage to leave". Collectively, the new community starts reshaping an alternative rewards system for the organization that can be external but more often is internal.

Cultural Formation

An organization's culture is created by its members' patterns of relating and communicating [5, 9, 17, 18]. As social beings, individuals watch and emulate patterns of behaviors within an organization, so they might "fit in" and be accepted. Impersonal and dehumanizing patterns of relating often are the norm within academic medical centers and therefore it takes a conscious effort to alter the "usual" patterns. A courageous person may set in play a new style of communicating that is experienced by others as more personal, more humane, or more life-giving, prompting others to emulate the same style. Similarly, what community members choose to focus on creates the new cultural values. Some organizations have adopted strategies of using positive deviance and positive psychological processes finding that discovering and focusing on what is positive and working well within their environment generates energy, enthusiasm, and a desire to propagate successes rather than focusing on the unending spiral of fixing what is not working [12]. When the patterns within an organization shift toward upholding the values and gifts of each member and keep a mindful focus on what is working well, the resultant "social-anthropologic artifacts" created by this new culture also shift. New patterns of communicating and relating are reflected in the minutes of meetings, in honors and awards bestowed upon worthy employees/students (e.g., Gold Humanism Honor society), the format and process for bringing new members into the organization (e.g., signing of an Honor Code), visible symbols of what the organization values (wearing an Honor Code pin on one's lapel), the work products that reflect collaborative efforts be it in clinical, education, or research missions, and booklets of compiled stories, poems, creative works of community members are created as gifts to incoming medical students, residents, and faculty. Such cultural artifacts are viewed, experienced, analyzed, and emulated by current and future members of the organization in a living process of cultural formation.

Evolutionary Process of Cultural Formation at IUSM

[T]he consistently most gratifying component of my responsibilities and day to day tasks has been our efforts at culture change and professionalism. We have ambitious goals in research and in clinical care, but I have publicly stated that if we exceed those goals but fall short of our efforts in professionalism, we will have earned a very hollow victory. Many of the stories we have collected are in a file on my computer. I often open that file and read a few stories. Doing so always relaxes me; with some I shed a tear owing to the pure emotion of what represents the very best in medicine.

(Professor of Medicine; Dean IUSM [7])

IUSM faculty, staff, and trainees have developed, led, and participated in a variety of activities, programs, retreats, and workshops to cultivate relationship-centered interactions and leadership skills, self-reflection and mindfulness in personal interactions, generosity of spirit, and attention to healing arts, all foundational abilities for achieving excellence in clinical care, teaching, and research while concurrently fostering cultural change within our work environments and ultimately faculty well-being.

Relationship-Centered Care Initiative/Discovery Team

IUSM began a three-year process of self-study and organizational development known as the RCCI in 2003. Funded by the Fetzer Institute and rooted in the observation made in the Pew-Fetzer Task Force that healing relationships are the foundation of humane, effective medical care [19], the RCCI's initial goal was to transform the culture of the School of Medicine by encouraging health-enhancing relationships in its social and clinical practice environments. The continuing aim of the initiative is to foster the development and conscious modeling of the empathy, thoughtful inquiry, and positive, respectful, collaborative interactions that we desire our graduates, faculty, and all members of our organization to exhibit.

One of the first RCCI efforts to create organizational change was a series of appreciative interviews conducted by the RCCI Discovery Team [15]. The Discovery Team included a small cadre of caring and committed IUSM faculty members who sought to manifest their personal values in their daily work. Appreciative Inquiry (AI) is a theory of organizational change that seeks to improve organizational culture by intentionally focusing attention on what is already working well in an organization and working to build on that foundation of existing success [12]. It suggests that by changing our focus of attention, we change our organization. We have begun to redirect the focus of attention and conversations away from "problems" and "barriers" toward what is working well in our organization and toward replicating or building on those successes we are already achieving.

The Discovery Team collected and analyzed over 80 interview stories that illustrate what is working right at IUSM. Because the good had never been an intentional focus before now, many, if not all, of those stories had never been heard or discussed. As a positive and hopeful image of IUSM began to emerge, more volunteers stepped forward offering to bring RCCI to their departments, committees, offices, or projects, stimulating a "rippling" impact throughout the organization.

The initial AI process stimulated interest in further work to develop IUSM community members' relational abilities and improve our work environments. Over time the number of faculty, staff, and students involved in cultural change activities began to grow significantly, from fewer than a dozen to over 160.

One IUSM faculty member from a regional campus, and a Discovery Team member, describes benefits in increased knowledge and skills, improved processes in meetings, and the development of relationships:

> *I was fascinated with the format of creating open, welcoming space for ideas to emerge and new connections to be made. It was amazing to see the model in action. I loved the rituals of checking in, appreciative inquiry, and appreciative debriefing. Discovery Team introduced me not only to new ideas but to many important new relationships with people at IUSM whom I had never encountered previously.*

(Volunteer Clinical Assistant Professor of Medicine, IUSM-Regional Campus)

The programs mentioned below describe the second wave of cultural change that emerged from the interest, desires, and needs of those engaged in the initial phase of the RCCI. Many of these programs are now provided as regular professional development offerings at IUSM.

The Courage to Lead Program

A year-long *Courage to Lead* (CTL) program was initiated in the Fall 2003, in response to participant requests for additional leadership development opportunities. Based on Parker Palmer's *Courage to Teach* program [4], the IUSM series consists of four 1.5-day sessions thematically linked to the four seasons: Fall, *Seeds of True Self*; Winter, *Darkness and Dormancy*; Spring, *Embracing Paradox*; Summer, *Coming into Fullness of Self and Community*. These retreats are designed to foster a process of personal formation for participants that enables participants to deepen their capacities to bring relationship-centered activities into the medical school. Participants learn how to create evocative and trustworthy spaces for learning and reflection, and bring greater depth to the work they are already doing. Formation work is framed by questions such as: "How can I engage in leadership roles and be true to myself?" "In what ways are my inner life and outer work connected?" What are the forces that nourish or constrain my soul?" "How can I find joy and fulfillment in my leadership roles and activities?"

CTL retreats provide time and support to reflect on the connection between attending to the inner dimension of our lives and the relationship-centered values and practices we wish to live by and extend to the entire medical school. In large group, small group, and solitary settings, the sessions explore the "heart of the leader" through stories from our own journeys, reflections on work experiences, and insights from poets, storytellers, and diverse wisdom traditions. Participants are invited to speak honestly about their lives and work, and to listen and respond to each other with compassion.

The impact of formation work is profound. As one participant describes it, this personal work is often life-changing:

> *This retreat series, through its self-reflection and personal sharing format in groups of 2's and 3's, allowed me to meet and deeply connect with some of my colleagues in such a unique way that is unparalleled when compared to all the other faculty development workshops or retreats that I've attended. Almost everything about that retreat was very conducive to enable the self-reflections and interpersonal connections: the isolated place, the safe emotional space, the pairing of people to share their stories, the open and honest questions, the skilled facilitators... I can keep going.... The uniqueness of this particular retreat is how restorative and yet powerful it is for the participants. Most other faculty development workshops/retreats are very task-oriented and want their participants to learn and practice new/specific skills which they will hopefully implement into their work places, but they don't usually enable participants to arrive at that deeper place within themselves to fully experience that restorative power like the CTL retreat.*

(Assistant Professor of Clinical Medicine; Education Leader)

Built into the retreats are the discernment of principles and practices that underlie Courage work, and an exploration of how participants might use these in furthering relationship-centered work at IUSM. The program fosters self-knowledge and the capacity for authentic presence and healthy relationships, helping participants to, in the words of Mahatma Gandhi, "be the change you want to see in the world." To date, more than 50 individuals have participated in this program that is offered annually.

Internal Change Agent Program

Thirty members of the campus community identified as potential change agents during the first three years of the RCCI enrolled in an intensive, five-session, 20-hour Internal Change Agent Program (ICAP). Facilitated by six members of the RCCI leadership team, ICAP was designed to help participants master knowledge and abilities that form the basis of successful culture change work. The sessions addressed key questions such as: What is an organization? How do organizations stay the same or develop new ways of being? And, how can we bring relationship-centered practices into organizational life (including sessions on dealing with difference, having difficult conversations, and using appreciative inquiry)? The ICAP series helped foster participants' personal awareness, facilitation skills, and ability to foster positive change in their local work environments.

Generosity of Spirit Conference

To expand the scope of the personal formation work that was started with CTL and ICAP, an IUSM conference based on the Fetzer Institute *Generosity of Spirit Project* was implemented in September 2006. Generosity—our ability to offer the best of who we are and what we have for the benefit of one another—is a valuable human attribute. In a world where media reports constantly remind us of conflicts, scandals, danger, and economic hardship, it can be easy to overlook individual and collective generosity. When basic generosity is not fostered, fear, mistrust, isolation, and self-centeredness can quickly infect the workplace, family, or community. Thirty participants from IUSM (health care providers, social workers, clergy, medical educators, and staff) participated in a 1.5-day workshop that used a curriculum of readings, poems, and oral story telling to examine the nature of generosity, conditions that foster or constrain generosity of spirit, and uncover individual gifts for and expressions of generosity.

Clarian Values Professional Development Program

Two national RCCI Immersion Conferences, one in 2006 and a second in 2007, were held at IUSM to share and promote work in medical school culture change. Teams from 16 academic medical centers from around the country have joined us to hear about the RCCI, observe IUSM's culture during participant observation immersion experiences, and to discuss culture change initiatives begun at their own institutions. The powerful experience of these two national conferences inspired us to use the same approach on a local scale. Interdisciplinary teams of coworkers (composed of physicians, nurses, allied health professionals, staff, residents, and other health care trainees) from within seven different clinical microsystems in our own IUSM hospitals and academic units have completed a one-year Clarian Values Professional Development Program, October 2006–October 2007. Teams have participated in four quarterly gatherings hosted at four different academic health care venues, enabling the entire cadre of more than 50 participants to observe a diversity of work sites and learn about the opportunities and challenges to organizational culture change through direct experience. Each team worked on a self-defined project over the year-long program, guided by an RCCI faculty facilitator.

In their reviews, it became clear that participants found the experiences and methods introduced during the workshops to be effective in improving their professional environments. For example, a breast care physician comments that she learned the importance of intentional work to create a strong team that focuses on improving interpersonal relationships, as well as a strong program:

The care of breast cancer patients requires a team of caregivers. As medical director, I had worked to build our program with a primary focus on patient care. The Professionalism Project helped me realize that I should focus on developing the team as well as the program—and it provided me with some practical tools that have helped tremendously. ... We have started a project consisting primarily of appreciative inquiry interviews with providers who interface with our patients. ... The dedication and compassion of each provider we have interviewed has been an inspiration to me. It has been educational to see the care we deliver from their perspectives. Inviting them to share their experience and insight with us has also served to identify them as part of this team. The recognition that we value their contribution has opened new communication pathways between providers. It has created a real sense of ownership for the success of our program among providers... [We have] more of a cohesive identity now where before it was a more of a collection of caring and skilled providers.

(Assistant Professor of Clinical Surgery)

A second participant describes increased energy and enthusiasm as a result of working with a team of similarly interested individuals:

The Clarian Professionalism Project is not even "complete" yet, but I have already seen benefits. I have developed friendships and strengthened business partnerships during my time on this project, and I hope to see tangible benefits to the patients at University Hospital very soon. If nothing else, it has shown me that sheer energy and enthusiasm (not only mine but also that of other team members!) can often achieve great things.

(Assistant Professor of Clinical Medicine)

Fostering Humanism

This program brings together story-based reflection on IUSM's professional culture, self-reflection, development of practical skills to address difficult professional situations, and cultivation of capacities for generosity and healing, to create a program that enables participants to see their professional environment with "new eyes", and to "be and act" differently in their everyday work and interactions. This unique interdisciplinary program brings together health care professions from diverse fields, departments, and schools to experience working with poets, artists, musicians, and writers to more deeply understand themselves and their roles as healers and caregivers. Participants listen to, reflect upon, and personally react to narratives, poems, music, and art. They also experience the creative expression of self by participating in small group sessions creating watercolor and mosaic art forms and/or writing music, poems, and narratives working with the Cancer Center art and music therapists and with local poets and writers. Participants have opportunities to meet together to continue as a poetry, writing, art, or music group between the longer 1.5 days retreats.

Leadership in Academic Medicine Program

Leadership in Academic Medicine Program (LAMP) is a monthly series of half-day workshops for new faculty in the School of Medicine. This program makes it clear to new members of the academic community that they are valued and that their success matters. The primary purpose of this program is to provide new faculty with both an orientation to the culture of the School and generic competencies that can help each of them succeed in their academic career. The program contributes directly to building a community where respect is a dominant, core value. Participants in LAMP form a cohort that remains together throughout the year building relationships across departments and disciplinary boundaries, and establishing a peer support network. In addition, each LAMP session concludes with a group-mentoring activity. LAMP participants are divided into groups of five, each facilitated by a member of the LAMP faculty, and are encouraged to discuss the challenges and joys they are experiencing as new members of the organization.

Leading Change in Academic Medicine

This program was founded on the belief that leadership skill development for our next generation of academic leaders is no longer an option; it is an imperative. The IUSM Leading Change in Academic Medicine (LCAM) program was designed for those who are, or will likely soon be, leading an academic enterprise, such as a division, department, or center. The curriculum is based upon the five leadership practices identified by Kouzes and Posner [20]: model the way; inspire a shared vision; challenge the process; enable others to act, and encourage the heart. These five practices were considered highly congruent with the orientation to a relationship-centered culture. LCAM sessions occurred monthly and were a half-day in length. LCAM participants formed a cohort that remained intact throughout the year, affording participants an overarching peer leadership network that bridged the many clinical and research divisions that make up our academic medical center. In addition, the larger LCAM cohort was divided into small teams that worked on a proposal for addressing an institutional challenge. These work teams were instrumental in building relationships among team members.

LCAM inspired participants to see and interact with the culture of the School:

> *I gained much that I was not expecting from the initiatives of which I was a participant. I went into the LCAM course hoping to learn additional leadership skills. While I certainly achieved that, I think it even more important that LCAM gave me an opportunity to broaden my view of the culture and climate of IUSM. I had a chance to partner with not only clinicians like myself but also basic science researchers and hospital administrators on a project. While we did not see a huge impact with our project, I personally benefited from the opportunity of seeing challenges faced by others outside of my own department and now have a better appreciation of what it takes to achieve the mission of IUSM. It is*

*helpful to see others working toward the same excellence for which I strive and gives me
hope that we will someday "change the world."*

(Assistant Professor of Clinical Medicine)

Effects of Cultural Change Activities on Faculty Vitality and Well-Being

Measuring the effects of organizational culture change is challenging. While the traditional, randomized, controlled research design would be ideal to use, culture change is neither linear nor static. Clarity about what faculty value and an emergent pattern of narratives in the direction of these values is one strong indicator of cultural change.

In order to systematically assess the impact of RCCI and related programs on our culture, in August 2007, a request was sent to all IUSM faculty, who had participated in one or more programs. Faculty were asked to check off which program/activity(s) they had participated in from a list including: Discovery Team meetings; Courage to Lead retreats; Internal Change Agent Program; Leadership in Academic Medicine Program; Leading Change in Academic Medicine; Teaching Caring Attitudes Workshops; Faculty Enrichment and Education Development Programs; Appreciative Interviews; RCCI Open Forums; and an "other" category. They were then asked to record their reflections on the following question: *"What effect (either positive or negative) did the program/activity(s) have on your energy, vitality and well-being as an IUSM faculty member?"* They were encouraged to write as little or as much as they liked and to respond for any or all of the programs/activities in which they had participated. Write-ins in the "other" category included: the RCCI National Immersion Conference; Clarian Values Professionalism Project; Competency Learning Community meetings; the Generosity of Spirit retreat; involvement in the RCCI newsletter editorial group; and the Gold Humanism Honor Society Selection committee. Over a one-month period, more than 30 stories ranging from a paragraph to several pages were received, almost all positive. The majority of respondents commented on the personal impact of the *Courage to Lead* seasonal retreats, but many also referred to other culture change activities.

Three major themes emerged from the faculty responses centering around coming to know oneself more deeply, the importance of finding and working with colleagues with whom one shares values, and creating a work-world where ideas, successes, failures, and efforts are valued and rewarded individually and collectively. Within each theme the stories could be categorized as reflecting personal insights, perceived impact, or observed outcomes. The analytic scheme and representative quotes from faculty stories are included in Tables 14.1–14.3 and are also described below.

Table 14.1 Personal formation

Knowing self: select quotes from IUSM faculty's stories about RCCI-related programs and well-being

Insight

[R]eal change if it is to last must happen within the individual by changing or shaping his or her world view. … By participating in numerous RCCI activities year after year many of the things I desired to do or be were continually reinforced in a very supportive and encouraging manner. The result was, I believe, an ability to achieve and sustain a new level of mindfulness. … Fewer meetings are onerous and my energy level is higher and stress level lower. …This is uplifting and reinforces the value of looking for the good that surrounds me. While these may not seem earth shattering, in the busy and often chaotic world of academic medicine, mindfulness, reduced stress, and more frequent positive encounters add up to a subtle, yet enormous difference. (Professor of Family Medicine; Executive Associate Dean)

If I am brutally honest with myself, I sought a career in academe because I love to learn. …, the most difficult learning for me is about self—self as I know it and self as others construct it for me. (Professor of Medicine; Associate Dean)

Perceived Impact

[T]hese programs have taught me to be more introspective and cognizant of my actions and their effects on our system at large. (Associate Professor of Clinical Obstetrics and Gynecology)

The RCCI programs have given me the opportunity to develop deeper self knowledge and mindfulness of the impact that my being and actions have on those around me, while working in an office that intentionally seeks to promote relationship-centeredness has allowed me to more fully live my personal values in my professional life. (Professional Staff in Medical Education)

When I look back over my professional career, I won't likely say I wish I had published more or had been awarded more grants or seen more patients on a half day clinic session or worked with more medical students but rather that I relished the excitement and passion of discovery, learning, healing through caring and being cared for by colleagues, students, patients, family and friends and knowing what I bring/brought to these dynamic and ever-changing relationships. (Professor of Medicine; Associate Dean)

The major effect it has had on my work has been to invite me to look at things differently and engage in dialogue about issues of concern. (Assistant Professor, Part-Time, of Psychiatry)

Interestingly, acting like relationships don't matter seems odd (when just a few short years ago that was the norm). (Associate Professor, Part-Time, of Surgery; Assistant Dean, Associate Professor of Clinical Medicine and Molecular Genetics, Adjunct Lecturer in Clinical Psychiatry)

I realized this was all about the other person and I found the strength to stay true to my own values and self. I see myself as someone who can remain calm, reflective, and comforting in times of stress. After the fact, I know I needed to return to this place of calm for ME and let the other person own their disruptive, chaotic behavior. (Professor of Psychiatry)

I found the sessions to be extremely helpful and helped me to reflect on my feelings and look at how I was not attending to my own needs. (Professor of Medicine)

I had been raised to value both personal and professional aspects of my identity; I had accepted a paradigm of separate "personal and professional selves" since my graduate training. … found the experience meaningful in providing an anchor for my professional challenges. These retreats and other RCCI activities have helped to re-create a "mindful" space that now allows me to not remain divided. In challenging professional moments, I find myself emerging with outcomes from a place of compassion and respect. (Adjunct Associate Professor of Family Medicine)

The retreats also happened to coincide with a time in my career when I was asked to shift my focus and take on a new area. … the chance to reflect, write, and share with colleagues

(continued)

Table 14.1 (continued)

helped clarify and sharpen my thoughts about this new direction. (Assistant Professor of Clinical Family Medicine; Assistant Dean)

Observed Outcomes

I can say that I am happier and more satisfied with my own skin than I have ever been. Professor of Surgery

This sustained attention to relationship, to recognizing the self while honoring the thoughts and differences of others in all our interactions, is life-giving for me. (Professional Staff in Medical Education)

RCCI does not so necessarily teach me new behaviors and ways of relating as much as it "frees" me to be my best self. (Professional Staff in Medical Education)

Table 14.2 Community formation

Finding community: select quotes from IUSM faculty's stories about RCCI-related programs and well-being

Insights

[W]e got to know more people and got to know them on a deeper personal level … people who seemed "off limits" before don't seem that way anymore, they seem just like people. (Associate Professor, Part-Time, of Surgery; Assistant Dean; Associate Professor of Clinical Medicine and Molecular Genetics, Adjunct Lecturer in Clinical Psychiatry)

Many of the academic groups that I now work with are on a first name basis. This is just a small thing, but it feels more comfortable to be in an environment where the attempt is being made to be less conscious of titles and hierarchy and show respect to all. (Professional Staff in Medical Education)

[T]he greatest benefit to me is that I discovered like-minded people and had an opportunity to explore or discern what I am being called to do with the next phase of my life in academic medicine. (Associate Professor of Medicine)

I forged some interpersonal bonds that are of mutual comfort in an ongoing way. I got to know members of the educational leadership teams on which I serve, and to understand their viewpoints on a deeper level than is possible through routine meetings on campus. (Associate Professor of Microbiology and Immunology, IUSM-Terre Haute; Education Leader)

[H]earing what other faculty were experiencing in their departments helped me realize that the hierarchical, non-collaborative style of the leaders in my unit was not the norm. After years of trying to fit in and make it work, I realized I couldn't change my work environment enough to ever be fulfilled … so one might say that the RCCI gave me the 'courage to leave.' (Associate Professor-anonymous department)

Perceived Impact

Through the RCCI/professional development programs, I acquired a better understanding of not only IUSM, but also my colleagues who care about serving and advancing the mission of IUSM. These experiences strengthened my self-understanding and how it relates to the demands of my work at IUSM as well as the need to meet them with dedication and compassion. The RCCI/professional development programs enhanced my appreciation of the need to be more explicit in aligning my professional values and their expression in my work. The result has been a stronger sense of my well-being as a member of the IUSM community. (Assistant Professor of Family Medicine)

Observed Outcomes

I was facing and dealing with some very challenging issues within my personal life. … While this issue loomed large, it was comforting to be surrounded by a caring and compassionate group of people and I felt lightened at the end of each day. (Adjunct Professor of Medicine)

Another benefit was getting to know some of the group's participants more intimately, understanding their own challenges such as career changes, difficult work situations, and

(continued)

Table 14.2 (continued)

frustrations. I also enjoyed the laughter and there was plenty of that. Closeness to some participants has continued beyond the workshops. (Adjunct Professor of Medicine)

Attending the DT meetings was energizing and provided me with a sense of hope and optimism. It was clear something different was happening at the IU School of Medicine. And I was envious of what was happening. I wanted it to happen in the School of Nursing too. (Professor and Associate Dean for Graduate Programs; School of Nursing)

The RCCI/professional development programs enhanced my appreciation of the need to be more explicit in aligning my professional values and their expression in my work. The result has been a stronger sense of my well-being as a member of the IUSM community. (Assistant Professor of Family Medicine)

Table 14.3 Cultural formation

Creating value: select quotes from IUSM faculty's stories about RCCI-related programs and well-being

Insights

I see RCCI as an intentional effort to create a work culture where the definition of professionalism includes acknowledging the humanity of colleagues. (Professional Staff in Medical Education)

[T]he check-in process consciously invites participants to draw upon both our personal and professional lives during the course of the meeting, two realms that are traditionally compartmentalized. … it makes an immediate difference who is in the room. … We collectively invoke a creative space where we can draw upon our personal and professional experience and knowledge. We are simultaneously individuals and part of something larger than ourselves, a powerful dynamic in which both the group and specific participants press us in a positive way to be our best selves. As we consider ideas and issues during the course of the meeting, we routinely discover nuances, dimensions and possibilities that nobody would have come up with alone. … Checking in allows us to surprise one another with details of our lives and then again with lively, engaging discussions where our faith in the process leads us to realizations that are as unexpected as they are incisive. (Professional Staff in Medical Education)

Perceived Impact

I was encouraged to share a model for developing relationship-centered care with first year students … and gained confidence that I might have something to offer not only my students but others in the medical community. (Volunteer Clinical Assistant Professor of Medicine; IUSM-Regional Campus)

We feel empowered to approach the administration with concerns because they embraced RCCI and made it clear that they are interested and willing to listen. (Associate Professor, Part-Time, of Surgery; Assistant Dean; Associate Professor of Clinical Medicine and Molecular Genetics, Adjunct Lecturer in Clinical Psychiatry)

We are not worried about squelching opposing voices, but welcome them as sources of new insight, even (or especially) if it requires re-thinking our direction. And most of all it means that we have permission to let go of the need to control the outcome of key initiatives and to trust in the transparent process of group deliberation. (Professional Staff in Medical Education)

I have connected with others even through disagreement and conflict to come out higher than where I started in understanding the very genuine goodness in who people are and what drives them in their work. (Associate Professor of Clinical Medicine; Assistant Dean; Education Leader)

(continued)

Table 14.3 (continued)

Observed Outcomes

I was given the opportunity to work with a team of professionals from many disciplines on the IUSM campus to develop a publication of creative works that included poetry, stories and visual art. ... Through this publication we were able to display the empathy, compassion and connectedness that we all long to feel and often fear is absent from medical culture. (Professional Staff in Medical Education)

We have brought basic science, clinical science and social science faculty together from across the campus to develop an innovative four year behavioral and social science integrated curriculum, and have found delightful partners ... developing processes that encourage faculty and staff to work together openly and honestly and that intentionally value difference has not only improved our work relationships, but has lead to much better work products as well. (Professional Staff in Medical Education)

At one point near the end of the meeting I sat back and observed the debate. I couldn't help thinking, "this is it. This is what relationship centered care is about." We came away with more than just an algorithm for triaging patients. We had come together as a group, we debated issues like a family, and we took a giant step as a team. (Assistant Professor of Clinical Surgery)

RCCI has also helped increase our productivity. ... In some ways it seems that our whole enterprise is a bit more collaborative. (Associate Professor, Part-Time, of Surgery; Assistant Dean; Associate Professor of Clinical Medicine and Molecular Genetics, Adjunct Lecturer in Clinical Psychiatry)

It was quite uplifting to witness the change in folks' openness and willingness to share and be vulnerable, all in the hope of making IUSM a "more livable space". ... The willing way in which so many faculty came forth in response to the RCCI efforts and shared stories of "doing good" and healing made me realize just how altruistic and compassionate most people in medicine ... are and can be. This realization boosted my confidence in my own work, in turn greatly improving my own well-being. (Project Manager RCCI)

Approximately, one year ago, largely due to a decrease in State appropriations, the IUSM faced the biggest budget crisis in our history. ... I believe that we, as a team, made some of the best decisions that could have been made under such difficult circumstances. ... I do believe that the RCC process we used to resolve the crisis has made us a far stronger institution. (Professor of Pediatrics; Executive Associate Dean for Research)

Knowing Self (Personal Formation)

Personal Insight

Faculty responses reinforced the fact that one of our most difficult challenges is to know oneself and that knowing oneself occurs in relationship with others. Staying true to "who we wish to be" requires vigilance and the kind of reinforcement promoted by RCCI activities to "sustain a new level of mindfulness." For some, RCCI activities have "freed" them to openly and more consistently express and live according to their deeply rooted values, behaviors, and ways of relating so they can "be their best selves."

Perceived Impact

Faculty expressed an increased appreciation of their role in co-constructing their reality at work and elsewhere. Faculty describe a growing realization that how they choose to speak and act in any particular situation and what they choose to focus on impacts their own well-being and the well-being of those around them, and of the organization in general. The power to influence conversations and exchanges in the moment by remaining mindful and centered is described as particularly useful during times of stress or chaos.

Observed Outcomes

Faculty members have noted that the mindful practice of staying aware of their internal compass has led to "liking oneself" and resultant happiness. Several faculty used the phrase "life-giving" in their stories suggesting that beyond the positive emotional impact of independent activities or experiences, individuals are finding energy, passion, and meaning in their broader experience of life.

Finding Community (Community Formation)

Personal Insights

As conversations about "what matters most" to individuals are encouraged, the "genuine authenticity" of individuals is opened up for others to hear and see. Titles, ranks, and other artificial institutional symbols of power become less important than the genuine open relationships between human beings. Partners in these new, ongoing relationships experience "mutual comfort" that results in deeper mutual understanding and increased clarity in personal decision making through perceived support. Even personal insights leading to difficult decisions, such as leaving one's current job and the organization when an unhealthy incongruence is uncovered, can be affirming.

Perceived Impact

Faculty members noted that the more they came to know and understand colleagues, the more clearly they could see and experience their dedication, strengths, and talents. In turn, understanding and appreciation of colleagues resulted in improved self-knowledge and heightened personal dedication, compassion, and need to stay true to one's personal and professional values. There was a tacit acknowledgement that in order to truly serve and advance the mission of IUSM, we must BE IN THIS TOGETHER.

Observed Outcomes

Several faculty spoke about their emotional experience of being a member of the IUSM community. Efforts to create a less-hierarchical system led to a more respectful and comfortable work environment. Difficult decisions around challenging issues were clearly made lighter when one felt surrounded by caring and compassionate colleagues. Laughter was a welcome and healing outgrowth of these deepening relationships. Individuals experienced a palpable sense of hope and optimism within IUSM that they wanted to share with other schools and within IU. As a result of participating in Relationship-Centered Care Initiative-related activities they were energized and felt a "stronger sense of well-being".

Creating Value (Cultural Formation)

Personal Insights

Rather than starting meetings with predetermined agendas, "checking-in" diffuses the sterility of the process by allowing individuals to share whatever they wish that will help them be fully present at the beginning of a gathering. Individuals listen to one another to better understand the states of mind of those in the room. Checking in humanizes the process by helping members appreciate and empathize with the complexity of colleagues' lives. Understanding, appreciation, and empathy generated from checking-in quickly establish a feeling of connectedness and improves communication since erroneous inferences and assumptions can be bypassed. For example, a person's yawn in the midst of a meeting's conversation is less likely to offend the other participants (who might feel the person is not interested in the conversation or ideas being discussed) if he/she shared at the start of the meeting that they had been up all night with a sick child. Creating the space and processes, which allow individuals to show up personally and professionally establishes new cultural norms/mores.

Perceived Impact

A culture that fosters understanding, appreciation, and empathy for its members allows quiet members of a group to feel as if they have something important to offer. Starting with listening rather than prescribed agendas has created a culture that is inviting and welcoming. Listening carefully to the voices of disagreement and conflict has also resulted in an appreciation of the "genuine goodness in people" to live out their values. Over and over the power of "trusting in the transparent process of group deliberations" resulted in better outcomes and greater satisfaction than a predetermined linear strategy.

Observed Outcomes

Faculty repeatedly reported increased productivity resulting from collaboration with colleagues and a keen recognition that the quantity and the quality of work products were greatly enhanced by the group process. The shared satisfaction experienced through collaborative work products seems to promote further connectedness, compassion, empathy, and the desire for continued collaboration. The positive experience of collaborative work was present across all missions: education, research, patient care, and administration. Even in the most difficult and fiscally challenging times, a more relationship-centered approach to administrative decision making reportedly helped strengthen the organization.

What Lessons from the IUSM Case Report Are Applicable to Other Institutions?

Keeping considerations of self and professional together permits us to see work as an expression of self, and professional aspirations for trustworthiness and virtuous action as aspirations of our own heart. In a field that demands as much of us as medicine, anything less than this integration of person and professional may be unsupportable in the long run.

(Thomas Inui, M.D.)

IUSM's story supports the idea that a more relational environment contributes to the psychological health of faculty and academic organizations, and that relationship with self is at the core of living an authentical and aligned or "undivided life". A focus on change from the inside out was one of our founding premises and instrumental in how we organized and interacted at every step. In this volume, McDaniels et al. (Chapter 5, xxx) also emphasize the importance of self-knowledge in the psychological health of faculty stating "*alignment between the individual and the system begins with self-knowledge. For example, a leader* (faculty member) *must believe within him- or herself that an ideal* (or value) *such as professionalism is not just a nice attribute to have within the organization. She must believe that the manner in which colleagues interact with patients, learners, and each other has a discernable impact on the entire mission of the organization—the quality of care, the mastery of teaching, the creativity and productivity of scholarship, and the wellbeing of all members of the unit.*"

All workers carry values into their efforts. In the case of faculty, these are a mix of personal values, general professional values, and values that carry them specifically into academic environments. Scholarship on personal and professional values in medicine suggests that there are a limited number of foundational values [3]. Research on general work values emphasizes the importance of similar values, such as altruism, being of service, respect, humanism, professionalism, citizenship, vitality, recognition, love of learning, and affirming a calling. Expressing these values in work requires an organizational environment that supports these values, promotes mindfulness, encourages employees to know and express self, enhances community, and advances individuals as the basis of value they create while embodying these values. The healthy worker, functioning at his/her best and finding meaning and value in work, is able to maximize these personal and professional qualities in the context of their work.

The professional development and cultural change programs described in this chapter have helped IUSM faculty to articulate and align their individual values with the values supported and upheld by IUSM's commitment (e.g., creating and distributing IUSM Core Values and Guiding Principles document, RCCI), new structures (e.g., recruitment of student and faculty willing to uphold IUSM Core Values, participation in culture minding faculty committees attentive to maintaining IUSM Core Values, input into IUSM leadership performance reviews that includes emphasis on the work environment) and ongoing processes and opportunities (e.g., participation in relationship-centered meeting practices such as "check-in" and nominal group processes) [21]. These programs provide some examples of how an organization can operationalize the model for supporting and sustaining faculty vitality and wellness described by Viggiano and Strobel in this volume (Chapter 6. Our faculty members' personal stories provide support for Viggiano and Strobel's statements: *"The institutional environment can be conducive to sustaining the vitality of individual faculty members. ... Most important, an institution's culture can help individuals sustain vitality by connecting individuals to the meaning and purpose of their work, expressing appreciation for each individual's contributions, and instilling a sense of community, or unity of purpose for accomplishing the institution's mission."* (Chapter 6, xxx).

At IUSM, we consciously worked to create trustworthy communities in many places, where people could show up fully and do their work in more authentic, relational, and collaborative ways. In some settings (CTL), participants had the chance to enhance their mindfulness, affirm their gifts, discern their limits, and expand the courage to "be the change"; in others (ICAP), participants honed their skills to be effective change agents. Sadly, there are still many academic medical centers where individuals feel devalued, dehumanized, isolated, and controlled. The good news, as our case study and the literature on relationship-centered administration and organizations suggest, is that a small number of vigilant, mindful individuals who pay close attention to day-to-day patterns of relating and conversing can produce significant positive change in the culture of large and complex organizations.

Organizational health and culture change are neither mysteries nor out of our grasp. They are, rather, the collective expression of a common desire to be our best, most creative and productive selves at work and live our shared values. We are reminded daily that the time for change in health care is now. The evidence suggests that by choosing to focus on what is working well it is possible to create organizations, whose processes embody the changes we most need and want to see.

References

1. Carter, J. (2006) *Cross This Bridge at a Walk* Nicholasville: Wind Publications.
2. Kirsch, D.G. AAMC President's Address 2007: "Culture and the courage to change" (Accessed at www.aamc.org 1/5/08).
3. Inui, T.S. (2003) *A Flag in the Wind: Educating for Professionalism in Medicine* Washington, DC: Association of American Medical Colleges.
4. Palmer, P.J. (1998) *The Courage to Teach: Exploring the Inner Landscape of a Teacher's Life*, 1st ed. San Francisco, CA: Jossey-Bass.
5. Stacey, R.D. (2001) *Complex Responsive Processes in Organizations* London: Routledge.
6. Litzelman, D.K., and Cottingham, A.H. (2007) The new formal competency-based curriculum and informal curriculum at Indiana University School of Medicine: overview and five year analysis *Academic Medicine* **82**, 410–421.
7. Brater, D.C. (2007) Infusing professionalism into a school of medicine: perspectives from the Dean *Academic Medicine* **82**, 1094–1097.
8. Palmer, P.J. (2004) *A Hidden Wholeness: The Journey Toward an Undivided Life* San Francisco, CA: Jossey-Bass.
9. Stacey, R.D. (2000) *Strategic Management and Organizational Dynamics*, 3rd ed. Harlow: Financial Times Prentice-Hall.
10. Streatfield, P.J. (2001) *The Paradox of Control in Organizations* London: Routledge.
11. Ryan, R.M., and Deci, E.L. (2000) Self-determination theory and the facilitation of intrinsic motivation, social development, and well-being *American Psychologist* **55**, 68–78.
12. Cooperrider, D.L., and Srivasta, S. (1987) Appreciative inquiry in organizational life: research in organizational change and development *JAI Press*, **1**, 129–69.
13. Watkins, J.M., and Mohr, B.J. (2001) *Appreciative Inquiry: Change at the Speed of Imagination* San Francisco, CA: Jossey-Bass/Pfeiffer.
14. Suchman, A.L., Botelho, R.J., and Hinter-Walker, P. (1998) *Partnerships in Healthcare: Transforming Relational Process* Rochester: Rochester University Press.
15. Suchman, A.L., Williamson, P.R., Litzelman, D.L., Frankel, R.M., Mossbarger, D.L., Inui, T.S., et al. (2004) Towards an informal curriculum that teaches professionalism: Transforming the social environment of a medical school *Journal of General Internal Medicine* **19**, 501–504.
16. Suchman, A.L. (2006) An Annotated Bibliography on Relationship-Centered Administration. Rochester, New York: Relationship Centered Health Care. (Accessed at http://www.relationshipcenteredhc.com/artman/publish/category_list.php).
17. Suchman, A.L., and Williamson, P.R. (2004) Changing the culture of a medical school using appreciative inquiry and an emergent process *Appreciative Inquiry Practitioner* **May**, 22–25.
18. Suchman, A.L. (2006) A new theoretical foundation for relationship-centered care *Journal of General Internal Medicine* **21**, S40–S44.
19. Tresolini, C.P., and the Pew-Fetzer Task Force (1994) *Health Professions Education and Relationship-centered Care* San Francisco, CA: Pew Health Professions Commission.
20. Kouzes, J.M., and Posner, B.Z. (2002) *The Leadership Challenge*, 3rd ed. San Francisco, CA: Jossey-Bass.
21. Cottingham, A.H., Suchman, A.L., Litzelman, D.K., Frankel, R.M., Mossbarger, D.L., Williamson, P.R., Baldwin, D.C., and Inui, T.S. (2008) Enhancing the informal curriculum of a medical school: a case study in organizational culture change. *Journal of General Internal Medicine*. **23**(6):715–22, 2008 Jun.

Chapter 15
Conflict Resolution in an Academic Medical Center: The Ombuds Office

Anu Rao, Patricia A. Parker, and Walter F. Baile

Abstract Academic medical centers are complex organizational environments characterized by the independence of faculty and the presence of multiple and overlapping goals. Several factors including competition for resources, financial strains, leadership turnover, and a negative research funding climate, all contribute to an environment with significant potential for conflict by virtue of jealousies, misunderstandings, misinterpretations, dysfunctional relationships, and other communication problems. Because conflict can have negative consequences including faculty dissatisfaction, stress, and burnout, it is essential to have strategies for resolving or managing it. In this chapter, we define and discuss conflict and its dynamics, review common formal and informal strategies for managing conflicts in academic medical center settings, and present an approach developed at The University of Texas M. D. Anderson Cancer Center. This alternative approach for conflict management utilizes an organizational ombudsperson, who does not replace existing structures for conflict resolution, but rather supplements them. We review the services that the Ombuds Office provides and describe the characteristics of the individuals who came to the M. D. Anderson's Ombuds Office in its first year and the types of issues that were addressed.

Keywords Conflict resolution, ombudsperson, burnout, academic health centers

A. Rao
Director, M. D. Anderson Ombuds Office, The University of Texas M. D. Anderson Cancer Center (M. D. Anderson), Houston, Texas, USA
e-mail: arao@mdanderson.org

P.A. Parker
Department of Behavioral Science, M. D. Anderson, Houston, Texas, USA

W.F. Baile
Department of Behavioral Science, Program, and Interpersonal Communication and Relationship Enhancement (I*CARE), M. D. Anderson, Houston, Texas, USA

T.R. Cole et al. (eds.) *Faculty Health in Academic Medicine*,
© Humana Press, a part of Springer Science + Business Media, LLC 2009

Introduction: The Need for a New Approach to Conflict Resolution

In this chapter, we will describe the definition and development of conflict and how conflicts affect individuals and the institution at an academic medical center. We will review common formal and informal strategies for managing conflicts in academic medical center settings and present an approach developed at The University of Texas M. D. Anderson Cancer Center where we developed an alternative approach for conflict management using an organizational ombudsperson, who does not replace existing structures for conflict resolution, but rather supplements them.

At a macro level, academic medical centers and medical workplaces are complex organizational environments characterized by the independence of faculty and the presence of multiple and overlapping goals. Over a period of time, faculty accumulate accoutrements of power such as space, grants, resources, personnel, laboratories, awards, and representation on prestigious national committees. Space is often a highly charged issue. Chairs are overwhelmed with middle management responsibilities while carrying out their scholarship. Furthermore, competition for resources, financial strains, leadership turnover, and a negative research funding climate all contribute to an environment with significant potential for conflict by virtue of jealousies, misunderstandings, misinterpretations, dysfunctional relationships, and other communication problems [1]. In health care settings, the press of time caused by patient care and interdisciplinary members of the clinical or research team also adds to the complexity of the conflict.

Faculty productivity and satisfaction are affected by the ability of faculty to communicate effectively and resolve conflicts in a timely and effective manner [2]. On a micro level, conflicts that are not resolved affect relationships with colleagues as well as direct reports and, in some cases, can lead to loss in faculty morale and increased turnover.

The human tendency to avoid conflicts is ubiquitous, but since conflicts are dynamic, avoidance can lead to negative outcomes as conflicts tend to get bigger, or more complex, if nothing is done to manage them. Conflicts rarely dissipate as they tend to spread in the workplace in a way that polarizes the members of a work group. In some situations, members of the workgroup tip-toe around the problem or the people in conflict, while discussing little else behind the parties in conflict. This "elephant in the room" syndrome perpetuates a culture of avoidance and may even result in reprisal for the person who breaks the silence of avoidance which would be taboo in such a culture.

Definition of Conflict

Conflict is often defined as "An expressed difference, disagreement, or dispute among two or more people" [1]. Conflict is not a mere disagreement; it is frequently accompanied by negative feelings arising from misunderstandings, mistrust, feelings of alienation, not being understood, betrayal, or feeling unappreciated or being put down. Marcus et al. take the view that a "conflict may exist within one person and unless it is communicated to the person towards whom there are feelings of frustration, disappointment, or anger, there is no expressed conflict." However, it is arguable that communication is accomplished by various means, not by words alone: body language and other nonverbal conduct such as exclusion, neglect, and apparent disrespect can convey conflict. Therefore, to start a conflict, the other side needs to *know* that there is a problem. As Marcus et al. [1] noted, an individual's inner conflict, which is their own feeling of private ambivalence or indecision is accentuated when a decision or action must be taken: for example, a chair may suffer many sleepless nights before communicating a concern of poor performance or unacceptable behavior to a faculty member. Conflict is thus intertwined with communication, and the management or resolution of conflict is also dependent on fostering healthy communication. In the absence of clear communication, negative sentiments, feelings, and motives are attributed to the other party in a conflict, and these negative perceptions and judgments prevent a rational discussion of each others' needs and points of view.

The Anatomy of a Conflict

The development of a conflict is best illustrated by an example of a simple conflict: at a faculty meeting the department chair recognized professor A for his good behavior and forgot to mention that a prestigious national award was obtained by professor B. Later that week, professor B, who was hosting a famous scientist from the well-known Northern University, sponsored a dinner for research collaborators, including faculty from the department and from local universities, at which time he "forgot" to invite the chair to the dinner.

At the interpersonal level, conflict is created and exacerbated by perceptions of threat or negativity where either or both parties infer and attribute hostile intent to the other party. Dana [3] used the term "Retaliatory Cycle" to describe the dynamics of interpersonal conflict, with the following sequence of events (see Fig. 15.1):

1. The triggering event: In the conflict described above between the chair and professor B, the forgetfulness of the chairman was a triggering event for Professor B, as he felt devalued by his departmental chair when he was not recognized among departmental peers for obtaining a prestigious national award.
2. The perception of a threat: Inferring hostile intent (e.g., "She is trying to sabotage me."; "He did this to make me look evil."; or in this case "He doesn't value me."), feeling affronted, rebuffed, devalued, neglected, or other negative emotions that are provoked by the words or actions of another. This is the cognitive or "appraisal" component. The perception of threat includes feeling "threatened" or feeling vulnerable in not being recognized or valued.
3. Defensive anger: Mobilization of hostile anger to protect one's ego from the insult. This is the emotional aspect of conflict and often is experienced as indignation, resentment, or lack of caring.

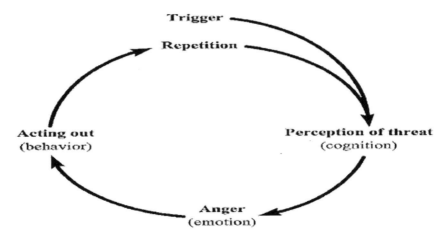

Fig. 15.1 The Retaliatory Cycle (Dana 2001)

4. Acting out phase: A behavioral component, a negative act, where the anger takes on action such as making disparaging statements, excluding someone from a group meeting or authorship of a paper. Passive–aggressive behavior is an important indirect way that others may express anger such as in the case mentioned above when the chair was not invited to the reception.
5. Repetition: In the example given above, the chair's forgetful behavior was perceived and interpreted by professor B as having a negative intent, and his anger was acted out in not inviting the chair to a departmental dinner, thus retaliating both by excluding him and disrespecting his role as a chair. If the chair then reacts by scolding the faculty member for not inviting him to such an important function the cycle of events can lead to a new cycle of trigger event, and a trajectory of reactions and behaviors of threat and acting out, thus creating a conflict of greater intensity. The feelings and behaviors in the repetitive phase are usually magnified in excess of the initial trigger event, and may lead to an escalation of feelings that can lead to a heightened level of conflict. The initial trigger event is replaced by a new trajectory of reactions and behaviors. This is one important argument for "Conflict Education" and the early intervention in conflicts. At the end of this chapter, we will discuss an alternative way to handle the rift between the chair and professor B.

Thus, the cycle of conflict has the perception of threat (the cognitive component), the feelings (emotional component), and the acting out (behavioral component). The effective resolution of a conflict depends on understanding the perception and "framing" of the events by the conflict participants and by asking questions to understand the sequence of issues, to separate the person (personality) from the problem (issues) and the dynamics of the problem (its evolution and escalation).

We prefer to use the term "Anatomy of a Conflict" to refer to Dana's description rather than retaliatory cycle [3]. In practice, the term retaliation is used to represent behavior that is a negative action or response to the action of others. For example, if an employee complained of sexual harassment, the harassing person may take some adverse action such as the denial of a promotion.

The example above illustrates interpersonal conflict. Conflicts between groups are called intergroup conflict and those that involve policies and culture occur at an institutional level.

Intergroup conflict can occur when group identity creates perceptions of threat, emotions, and behaviors that escalate the conflict. An example of an intergroup conflict is seen when subgroups in a workplace polarize according to professions: for example, doctors versus nurses or nurses versus pharmacists; or when one group clashes with another because of identity of race, ethnicity, or gender.

Institutional conflicts include concerns of policy, rewards, processes, and culture. Depending on the issue, the problems could be simple or complex. On occasion, the problem is intertwined with intergroup conflict and revolves around issues of employment, careers, power, management practices, and organizational culture.

Consequences of Conflict

Conflicts can transform interactions from simple disagreements of words, to rifts in relationships that manifest itself in a number of ways: lack of civility, abusive words, spreading of gossip and negativity, triangulation of a third person into the conflict, preferential treatment, and abuse of authority by retaliation. It is common to see each party in the conflict attribute negative motives to the behavior of the other party while judging itself by its intentions, tries to win, while portraying the other party as unfair and unworthy of support. Typically each party engages in a defensive maneuver that increases anger and anxiety of the other party and triangulates other members of the organization.

We contend that the stress associated with conflict and snarled communications between leader and led, or even colleagues, can lead to emotional turmoil, poor relationships, and loss in productivity [2]. Messy conflicts can create rifts, poor climate, and increase the turnover in an academic workplace [4].

In medical institutions conflict can occur at any level and contributes to the high rate of stress and burnout among medical school faculty, mid-level managers (chairs), and upper-level managers (deans) [5–9]. The more central the leader, the wider the ripple effects of conflicts. Deans who are preoccupied with conflict cannot lead. When department heads do not seize the moment to address and redirect conflict they miss the opportunity to mentor their faculty. Faculty who are preoccupied with conflict show less productivity and clinically have more dissatisfied patients, tend to make more medical errors, and feel more burntout [10]. When conflicts are not managed they can spiral out of control, creating a crisis when suppressed and avoided they fester and can *lead to a culture* of conflict, avoidance, and denial, which becomes woven into the fabric of a department. Careers can be dislodged, and for some they are destroyed.

People in power may use institutional resources and legally defensible methods to buttress their position, which is seen as being unfair by the party in opposition. When the conflict is poised to have a winner and a loser, the loser fears retaliation and alleges that those in a position of power will retaliate against them, and thus escalate the conflict. Since conflict can have many diverse and far-reaching consequences, both formal and informal strategies for resolving and managing conflict are essential for an organization to function effectively.

Formal Resources for Conflict Resolution

Institutions have both formal and informal resources to handle conflicts. We briefly describe some of these formal and informal resources.

1. Human Resources

Institutions tend to rely on formal systems such as the Office of Human Resources (HR) for resolving conflicts. These systems are often seen by employees as being partial to the management. Indeed, HR offices have a formal role in organizations as they represent the institution in hiring and discipline matters, and provide for a uniform administration of institutional policy and practice.

2. The Chain of Command

Relying on management or the formal chain of command is another formal strategy that many academic institutions use to resolve conflict. Faculty members may discuss their concerns with the department chair or with senior faculty members to pursue their interests. This system works well for most faculty. However, there may be conflicts with the chair, beliefs that the chair may not be interested in getting involved, or there may be issues that present a perception of conflict of interest. In these circumstances, a conflict may not be addressed by either party. For example, a faculty member may feel that a performance evaluation is unfair, but say nothing about it for fear that it may have an impact on future promotions or references. He or she may decide that an argument about performance may jeopardize the relationship or affect letters for promotion, or references. A professor can create a research climate characterized by conflict by creating perceptions of favoritism, by dictatorial rule over junior faculty or fellows, or by creating an atmosphere where underlings compete for rather than collaboratively utilize resources. This climate is not conducive to raising questions about a hypothesis or a failed experiment; in such a situation, there may not be a legal problem but precious institutional resources are wasted and careers are adversely affected.

3. Appeals Process

Typically, organizations have appeal processes to handle employment decisions such as nonrenewal of contract. The formal Appeals Process is designed to handle employment decisions between employer and employee, but does not address conflict, or the morale and behavioral problems caused by poor conflict management, such as lack of respect, competitiveness, misunderstandings, and lack of trust. Formal systems do not address the need of faculty members to have a resolution that manages the relationship while developing a solution, as in the case of a faculty member who has a conflict with regard to authorship or one who is frustrated in attempts at collaboration in research.

A formal problem-solving approach is usually set up in an arbitration model, where two or more parties present their problem to a person in authority, a neutral third party who makes a decision. This approach resolves the problem but sanctions one party. Rather than solving the problem, this approach determines who wins and who loses and may polarize the people in the dispute.

Informal Resources for Conflict Resolution

1. Human Resources Consultants

Sophisticated and resource-rich organizations use consultants to solve group and management conflicts in an informal mode. Management consultants are brought in ostensibly as outsiders to make an objective analysis of the situation. Work group members may not always view this as "objective" as management consultants are sometimes viewed as being tools of the manager or leader.

2. Employee Assistance Programs (EAP) and Faculty Assistance Programs (FAP)

Faculty members face the pressures of work as well as personal and family stress. Additionally, faculty may have behavioral or psychological problems, which can be helped by having an EAP or an FAP. These personal and psychological concerns are different from workplace/work environment concerns.

3. Faculty Development, Coaching, and Training

Organizations have progressive approaches to regulate the way in which people relate to each other at work. For example, M. D. Anderson implemented an innovative Faculty Leadership Academy and many programs to train faculty in communication, mentoring, and interpersonal skills. Faculty leaders have opportunities to receive coaching and attend educational programs at many locations.

Organizational strategies to deal with change and integration include climate surveys to understand the extent of the problem, and leaders are encouraged to develop emotional intelligence and develop change management strategies. This chapter argues that simultaneous with organizational education and change efforts, complex organizational environments need individual and localized strategies to help people navigate the turbulent waters of conflict and change in a way that helps them save face, maintain their professional reputation, and operate on a local level so as to minimize the negative effects.

Organizational investments in academic management and leadership have included training for faculty (Faculty Leadership Academy), management training, coaching of individual faculty members, management interventions, and redesign of departments. These efforts are mostly proactive and in some cases they are educational and preventive. In many organizations the management interventions and counseling services operate as parallel services, each bound by its own rules for confidentiality. The M. D. Anderson faculty leadership development program, for example, has resulted in chairs initiating full team alignments in their departments. Team alignments of groups of faculty and staff frequently involve dealing with unresolved conflict that may have festered for years, damaging the team's ability to work together to achieve their goals. Using conflict resolution techniques, group members safely air their differences and learn to discuss, disagree without rancor or blame, and negotiate resolution. The results for some teams have been remarkable as old differences are laid to rest and the team is freed up to move forward.

If one maps existing formal and informal strategies to manage conflicts, it becomes apparent that what is missing for faculty in many institutions is a confidential setting where they can get help in solving awkward problems, difficult situations, or working with difficult people. This gap in institutional resources is closed by the creation of an Ombuds Office, which provides a unique solution to the cross section of problems of individual and group life in the organization. It also gives people a voice and helps them to feel that there is at least one place to which they can go, where someone who is designated as neutral by the institution will listen to their perspective. This listening and, in some cases, the reconciling of divergent perspectives by an office, which is institutionally neutral to departments, rank of faculty, or other characteristics is exactly what is needed. In addition, an internal neutral office, which understands the inner workings and culture of the organization helps both leaders and faculty in negotiating difficult moments in their career.

Ombuds Office: An Innovation for Conflict Management

Background

An Ombuds Office is an innovation designed to assist in all stages of problem solving, as well as to be the office of last resort. It is a confidential, neutral, independent, and "off the record" office where a faculty member can talk without worrying about legal repercussions, or faculty can obtain consultation and assistance with problems, in a way that makes common sense to the individuals or parties in dispute. Problems that are brought to the Ombuds Office range from issues of respect to problems that may involve institutional policies.

To assist faculty with constructive approaches to conflict, obtain mutually acceptable outcomes, and reduce lawsuits and organizational damage, many academic and corporate organizations have created an organizational entity called the Ombudsman.

The Ombudsman is designated by an institution as a neutral office and is a modified version of the classical ombudsman role, which originated in Sweden at the turn of the 18th century. The Office was created as a protected platform to assist everyday citizens from the power of feudal landlords and industrialists, to restore rights of personal property. The Ombudsman used both mediatory and arbitrative approaches to drive decisions. This model of Ombudsman as a final arbiter of justice is called a "Classical Ombudsman." Many state legislatures have implemented the classical Ombudsman model to provide mediation and alternative dispute resolution approaches as a way to reduce employment disputes and litigation.

Many US and international corporations including the Federal Government have modified the Classical Ombudsman role to an "Organizational Ombudsman," who operates according to the standards of practice of the International Ombudsman Association. The Standards of Practice are also accepted by the American Bar Association and include four foundational principles of Confidentiality, Impartiality, Independence, and Informality/off the record. The Ombudsman is a designated institutional neutral, who provides a safe and confidential space for faculty to discuss concerns and problems. The independence is assured by freedom from conflict of interest in the chain of command and bureaucracy. In addition, the Office does not represent the institution as it does not receive notice on behalf of the institution.

The Ombudsman typically reports to the chief executive officer or the president of the institution. While seemingly a paradox that the Ombuds Office gains independence and credibility by reporting to the highest administrator of the Institution (usually the President or CEO), it actually assures a "neutral space" for operating outside the confines of the normal bureaucracy, whereby the Ombudsman is not affected by the special interests of one level or segment of the bureaucracy over another. Faculty are concerned that they must remain independent of the administration or the clinical management system, therefore this organizational arrangement is acceptable to the faculty. In addition, reporting to the President or CEO provides symbolic authority and legitimacy for the Ombuds Office. The International Ombudsman Association's Best Practices also underscores the importance of such an organizational arrangement.

Establishment of the Ombuds Office at M. D. Anderson Cancer Center

Dr. John Mendelsohn became President of M. D. Anderson in 1996. At Dr. Mendelsohn's request, a Committee on Conflict Resolution was formed, and asked to recommend a process for faculty conflict resolution. Dr. William Brock, an active member of the committee and a professor in the Department of Experimental Radiation Oncology researched conflict resolution systems in US academic medical centers and proposed the establishment of an Ombuds Office for faculty dispute resolution. Dr. Mendelsohn supported the committee's suggestion and the Faculty Ombuds

Office was established as part of a revised conflict management program for faculty. In 2003, this service was extended to the nursing staff and Ms. Janice Freeman, a well-respected member of the Nursing staff was selected to provide Ombuds services on a part-time basis. Her services were parallel to, but separate from the faculty Ombuds Office. In 2005, it was decided that the services of the Faculty and Nursing Ombuds Office needed to be expanded to all employees of the institution.

In 2006, Dr. Anu Rao was recruited to establish the M. D. Anderson Ombuds Office. Dr. Rao's experience included similar roles at Coca-Cola Enterprises and Princeton University; she was charged with expanding the Ombuds Office to cover all 17,000 employees as well as 4,000 trainees and fellows. The Faculty Ombuds Office was transformed into the M. D. Anderson Ombuds Office. The change of name to "Ombuds" was intentional in order to represent a gender neutral setting that was not oriented to the "male" gender term represented by the word "ombudsman." Additional staffing was needed and Dr. Rita Burns was recruited as a second full-time ombudsperson. Dr. Burns had previously worked at Marquette University as an Ombudsperson.

The M.D. Anderson Ombuds Office offers a range of services to M. D. Anderson employees, trainees, and fellows, all geared toward improving productivity, efficiency, and creativity in the workplace. Employees contact the office for information about policies, common institutional practices, or the availability of other employee services. Employees also seek guidance or strategic advice for handling virtually any work-related issue, especially those for more effectively dealing with supervisors and colleagues. The largest segment of the work involves facilitating problem solving between two or more individuals. For this, the Ombuds use a variety of techniques, including shuttle diplomacy, informal mediation, negotiating, and so forth.

The Ombuds Office differs from above-mentioned formal resources in that it offers a confidential, neutral, independent, and "off the record" services for faculty. These four characteristics work in synergy to provide a safe place to discuss concerns and work out solutions. The staff are trained to handle disputes in a way that does no harm. Additionally, the organization must create policies of non-retaliation that prevent backlash against those who use the office in good faith.

The Ombuds Office can work with organizational resources and programs: referrals of stressed faculty to the EAP, identify issues for training, development, and coaching, and provide preventive services to reduce appeals and legal claims.

How the Ombuds Office Functions and an Overview of Ombuds Services

The Ombuds Office functions with a short-term, problem-focused intervention approach. Most cases are solved within 1–4 weeks and a small number of cases reflect situations that involve difficult or intractable cases: usually those with systemic dimensions such as leadership or organizational problems. Faculty visitors to the Ombuds Office present concerns of poor relationships that may affect their careers or promotions. These may include salary inequities, performance evaluations, unfair practices, issues of authorship, research-related conflicts with fellow researchers or Principal Investigators, or concerns of sexual harassment and discrimination.

Individuals or groups who seek services at the Ombuds Office are referred to as "visitors" rather than clients, as the latter conveys a fiduciary relationship between the parties and assumes loyalty or duty to one party. Since the office is an impartial and neutral office, the term "visitor" signifies a functional relationship that is oriented to the problem-solving role of the Office.

The Ombuds Office provides the following concrete services to faculty visitors:

1. *Clarification of Issues and Perspectives*: The Ombudspersons actively listen to individuals and when needed (with permission) listen to the other party in dispute. Listening itself has value, both from a therapeutic and a fairness perspective. Ombudspersons engage in the problem discussion to clarify issues and gather the full context of the visitor's concern. They provide information about their role as neutrals with no decision-making authority, and clarify mutual expectations as to what the Ombuds Office can and cannot do. The visitor is asked what he/she has done to resolve the issues and what his/her objective is in this situation.

2. *Providing Information and Options*: Some visitors seek information about policies, resources, organizational practices, and sources of information. If the individual requires specific organizational information or interpretation of policy, the Ombuds will gather the information and, when appropriate, seek clarification from the policy steward. The bulk of the work, however, revolves around assisting individuals with work-related concerns that involve other employees. After roles and expectations are clarified, and the needs are discussed, there is discussion of options. Each option is reviewed, the pros and cons are evaluated, and a plan of action is discussed. Ultimately, each visitor decides on his/her own plan of action. They may choose to handle the situation themselves, use other M. D. Anderson resources, including Human Resources or the Employee Assistance Program, or they may choose coaching or some other intervention by the Ombuds Office.

3. *Coaching*: The term coaching is used to describe several related forms of supportive activities, including clarification of issues and developing a strategy for presenting them directly to other involved employees. This may be a face-to-face meeting or written communication. Coaching helps visitors clarify values, understand their role in the development and escalation of conflicts, consider consequences of inaction versus all possible actions, and, finally, discover new ways of approaching problems.

4. *Framing* of a conflict is also an important element in conflict management. Framing is defined as a subjective mechanism that tries to make sense of our experience. One person may try to figure out *what* the issue is, and *how* important

it is, while another person may focus on what it is they *want as an outcome*. The Ombuds helps the visitor to assess how important the issue is to him/her and to frame the issue in the most constructive manner by thinking about the problem from the other's perspective. A critical aspect of framing is that the "other party is generally characterized by the visitor in negative terms, as a 'wacko" or as "evil." In a neutral setting, the conflict is framed in more objective terms, as one's needs or interests, without demonizing the other party.

5. *Interventions On Behalf of the Visitor*: With the permission of the visitor, the Ombuds will assist in the management of conflicts by intervening as a neutral and impartial third party. Based on the issue at hand, the Ombuds can provide facilitation, mediation, shuttle diplomacy, or taking the issue forward to a higher level.

(a) *Facilitation/mediation*: A face-to-face negotiation between parties, with the Ombuds serving to moderate the conversation. This keeps the discussion focused on problem solving. The purpose of facilitation is to give the parties an opportunity to understand each other's concern and arrive at a solution that is mutually satisfactory.

(b) *Shuttle diplomacy*: Negotiation with the Ombuds talking to each party separately. This helps each disputing party understand and negotiate with one another, through a neutral third party, when face-to-face meetings would be too stressful.

(c) *Take issue forward*: In some cases, always with the permission of the visitor, the issue is taken up by the chain-of-command to the higher authority, frequently the chair, department director, or vice-president.

Effective conflict management is enabled by reciprocal communications of mutual needs and interests at a time when one's natural instincts lead parties in conflict to a flight/fight response, which turns into avoidance or aggressive actions.

Program Usage, Issues, and Concerns of Faculty and Clinicians

Who Came to the Ombuds Office?

During FY07, the Ombuds Office assisted a total of 427 visitors from various institutional positions and departments [11]. In 2007, 64 faculty members visited our office, distributed equally across rank and tenure, and years of service (see Fig. 15.2). The total numbers are consistent with other academic/medical Ombuds programs of our size. For example, The National Institutes of Health had 328 visitors in the first year of their Ombuds program [12]. At M. D. Anderson, the number of cases per month increased during the first year of the expanded program, with increasing awareness of our program. We expect the total number to increase further during FY08.

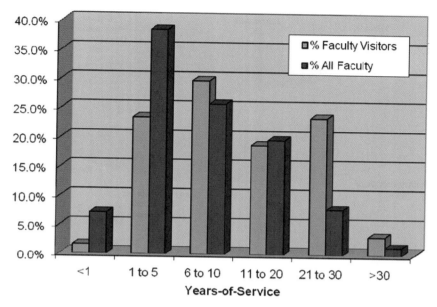

Fig. 15.2 Percent of faculty visitors who came to the Ombuds Office in FY07 by years of service

Diversity of Faculty Visitors

Of the 64 faculty members who visited the program in the first year, the majority were White. A small percentage of Asians and African Americans visited the office.

A few more women than men came to the Ombuds Office. Women faculty presented issues that were known to women in science and medicine. For example, some complained they were more likely to be evaluated on relationships with staff, while others alleged that they were not encouraged and coached the same way as men regarding tenure-track and promotion.

The Focus of Faculty Concerns: Relationships and Careers

Faculty members came to the Ombuds Office to discuss relationships with peers, faculty supervisors, departmental chairs, and division heads. As seen in Fig. 15.3, most of the concerns were centered on relational issues, but 30% of the concerns related to career concerns such as tenure, re-appointments, or the termination of appointments.

Most faculty members were concerned about the quality of the relationship or about troubling interactions. Only a quarter of the faculty complained about administrative actions. This is a propitious signal, as it indicates that most faculty members

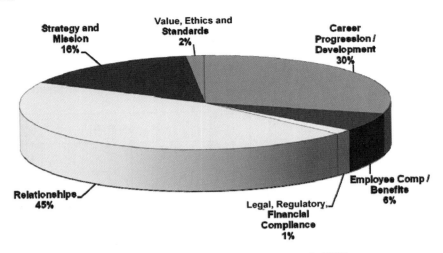

Fig. 15.3 Percent of faculty who raised concerns in each category in FY07

who visited the Ombuds Office were motivated to improve the quality of their inter-actions. This data has implications for coaching and training in conflict resolution.

Issues in Interaction: Collaboration, Civility, and Equity

As seen in Fig. 15.4, 40% of the faculty members who had concerns about relation-ships focused on the poor quality of collaboration with their peers. Twenty-two percent had concerns about equity and issues of fairness. Other concerns included civility, issues of respect, and communication. The concerns that faculty had with their peers focused on items such as civility, authorships, and agreements for laboratory or clinical collaborations. Civility concerns included disrespectful communications, rude behavior, or exclusion from communications or work products that violated previous understandings. Once again the data showed that faculty can gain from learning skills in communication, conflict management, and conduct related to civility.

Administrative Actions

Of the small number of individuals who face administrative actions, most were con-cerned about performance discussions, especially about the need to improve perform-ance. A few faculty members who had faced disciplinary actions sought redress through the Ombuds Office since they believed that they had been unfairly sanctioned by the Institution. The administrative actions included performance discussions, notice for nonrenewal of appointments, and notice of termination of appointments.

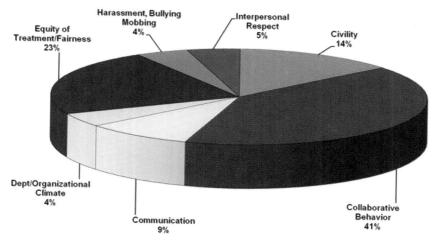

Fig. 15.4 Percent of faculty concerns regarding quality of interaction

The Faculty Conflict Resolution Policy invites faculty to seek the services of the Ombuds Office prior to seeking a formal appeal, and some faculty avail themselves of this option, as it provides a safe avenue for discussion prior to taking action.

Career Concerns and Mission-Related Concerns

With regard to career issues, faculty were concerned about issues of tenure and security of the position, while others were concerned about the career development, coaching, and mentoring for careers at this medical center.

As Fig. 15.3 shows, nearly equal numbers were concerned about the leadership and management of the medical center and the policies and procedures that pertained to faculty. The models of academic leadership may need more exploration and discussion, but it is outside the scope of this chapter.

The Focus of Faculty Concern: The Leadership

Among faculty, poor interactions with chairs and division heads was a common concern (see Fig. 15.5). It is obvious that the quality of this relationship is critical as it can have serious consequences. Negative interactions with a faculty supervisor may result in a lack of support for a research activity or a new clinical endeavor. Conflicts with the leadership may result in a loss of motivation, morale, and career development. Some leaders avoid discussion of problems with their faculty, preferring instead to defer the problem to their administrative staff to handle. This often

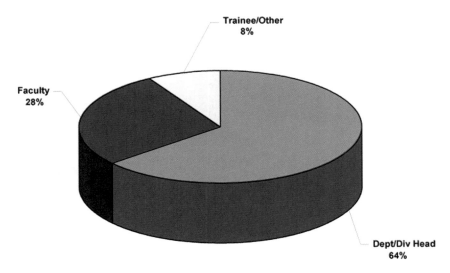

Fig. 15.5 Percent of faculty concerns regarding leadership

creates new tensions and conflicts. Such deflection of problems enlarges the scope of a problem and increases anxiety among all members of a faculty unit.

It is critical to coach faculty leaders to use conflict as a diagnostic tool to observe their communication style, issues of parity and fairness, jealousy among peers, and as an opportunity for mentoring and coaching of staff. Academic medical centers pose a special challenge as faculty leaders are pressured to divide their time between their own research activity, patient care, grants, maintaining visibility, and mentoring members of their group. Ordinary impatience can change the quality of an interaction from a nurturing discussion to a brusque encounter.

Lessons Learned: The Need for Discovery

The faculty members at M. D. Anderson are leaders and drivers of a system that produces and consumes knowledge, while providing the state-of-the-art patient care in cancer medicine. Given that our faculty and staff are focused on the mission of our cancer center, we are dedicated not just to the resolution of conflicts, but to share our learning to improve the conflict management competency of each faculty member (and all employees) as well as to explore the systemic dimensions of conflict.

In the first year of the program for all employees, we have learned that we need to pay close attention to the nature of conflicts in order to develop some hypotheses about career development, communications, mentoring, and its impact on the organizational culture.

Simultaneously, we need to look at the structure and efficacy of formal systems for conflict resolution and the incentive that is offered by the organization to resolve issues at the informal level. We will need to partner with an array of educators, trainers, coaches, and counselors at the Medical Center to study the systemic dimensions of conflict and the interventions needed.

Organizational culture plays an important role in establishing cultural cues for civility, collegiality, and conflict management. We need to understand more about the impact of perceptions; is a faculty chair seen as mentor or leader/manager? How does the relationship structure affect the ability to give and receive feedback on performance? What is the impact of organizational culture on the behaviors of dialogue and deference, or on expectations of compliance and followership?

Applications of the Ombuds Approach to Conflict

We return to the case of the conflict that was described earlier in the section, "The Anatomy of a Conflict," to demonstrate how the Ombuds Office might approach this issue. We will assume that the chair spoke to the Ombuds about escalating tensions between himself and Professor B.

The Ombuds clarifies the goal of the faculty member and understands that he genuinely forgot to recognize the award that Professor B received. Other faculty had talked to him about the dinner with the professor from Northern University. The Ombuds offers to talk to Professor B. As an option, the Ombuds will outline another possibility: here, the chair could initiate a discussion with Professor B. In the discussion, he would acknowledge his mistake, that he should have recognized Professor B at the faculty meeting, and apologize without going into a major confession. The chair will go one step further, and make amends for the mistake by having a departmental reception to honor Professor B for his achievement. What do you see as the outcome?

The lesson here is that it is the chair who models the way, and rises above the conflict to improve the relationship. By setting an example to Professor B, he teaches institutional values in a way that shows integrity and cultivates trustworthiness.

The Ombudsperson coaches the chair to display these leadership traits versus confronting Professor B with negative tactics. The teaching and learning is that these honest interpersonal strategies prevent an escalation of conflict and create new trajectories for their relationship

References

1. Marcus, L.J., Dorn, B.C., Kritek, P.B., Miller, V.G., and Wyatt, J.B. (1995) *Renegotiating Health Care—Resolving Conflict to Build Collaboration* San Francisco, CA: Jossey-Bass.
2. Hearn, J.C., and Anderson, M.S. (2002) Conflict in academic departments: an analysis of disputes over faculty promotion and tenure *Research in Higher Education* **43**, 503–529.
3. Dana, D. (2001) *Managing Differences: How to Build Better Relationships at Work and Home. 3rd ed.* Prairie Village, KS: MTI Publications.

4. Nettleman, M., and Schuster, B.L. (2007) Internal medicine department chairs: where they come from, why they leave, where they go *American Journal of Medicine* **120**, 186–190.
5. Gmelch, W.H., and Burns, J.S. (1993) The cost of academic leadership: department chair stress *Innovative Higher Education* **17**, 259–270.
6. Saleh, K.J., Quick, J.C., Conaway, M., Sime, W.E., Martin, W., Hurwitz, S., et al. (2007) The prevalence and severity of burnout among academic orthopaedic departmental leaders *Journal of Bone and Joint Surgery American Volume* **89**, 896–903.
7. Cruz, O.A., Pole, C.J., Thomas, S.M. (2007) Burnout in chairs of academic departments of ophthalmology *Ophthalmology* **114**, 2350–2355.
8. McPhillips, H.A., Stanton, B., Zuckerman, B., and Stapleton, F.B. (2007) Role of a pediatric department chair: factors leading to satisfaction and burnout *Journal of Pediatrics* **151**, 425–430.
9. Gabbe, S.G., Melville, J., Mandel, L., and Walker, E. (2002) Burnout in chairs of obstetrics and gynecology: diagnosis, treatment, and prevention *American Journal of Obstet Gynecol* **186**, 601–612.
10. Marshall, P., and Robson, R. (2005) Preventing and managing conflict: vital pieces in the patient safety puzzle *Healthcare Quarterly* **8**, 39–44.
11. The University of Texas M. D. Anderson Cancer Center Ombuds Office. M. D. Anderson Ombuds Office Annual Report: FY 2007.
12. National Institutes of Health. NIH Office of the Ombuds Annual Report 2000. http://www4.od.nih.gov/ccr/fy00annualrpt.pdf

Chapter 16
Preserving Principal: Programming for Faculty Health and Well-Being

Thelma Jean Goodrich

Abstract Programs in faculty health have elements that vary in scope and depth. Some elements focus on services for individuals, some on education for the whole faculty. Some focus on interventions for the impaired, others on prevention of impairment. This chapter describes these aspects of programs and their various formats. As a resource for new or expanding efforts, the review is meant to be suggestive rather than prescriptive and to support imagination and creativity available in every setting.

Keywords Programs, formats, seminars, groups, culture

Eighteen people in white coats sat at tables arranged in the favored U-shape. It was 7:30 in the morning. Most were eating bagels and fruit; a few were not awake enough to eat bagels and fruit. The gathering and its timing were routine for this department—a monthly faculty meeting.

Within 15 minutes, I completed my presentation on our institution's faculty health program. I listed all its benefits for the faculty: educational seminars on stress management, panels on work/life balance, responsiveness to requests from faculty, confidential counselors, meditation group, and renewal group. When I asked for questions, the eminent Chair of this eminent department responded in low and measured tones, "It's very nice of this institution to offer a program that represents care and concern for the faculty, but it's just a bandaid. Just a bandaid. The real problems are institutional, and if the institution really wants to help us, it can make some changes itself. One example is the bombardment of e-mails every day that have no bearing on our work here in this department. We don't need to know that Garage 6 way across campus will be closed Saturday. And then there are all those surveys. Plus all those required online courses. These have nothing to do with my work. If the administration really cares about us, it should back off so we

T.J. Goodrich
Director of Faculty Health, Department of Behavioral Science, The University of Texas M. D. Anderson Cancer Center, Houston, Texas, USA
e-mail: tjgoodrich@mdanderson.org

T.R. Cole et al. (eds.) *Faculty Health in Academic Medicine*,
© Humana Press, a part of Springer Science + Business Media, LLC 2009

would have a less stressful environment. Then we wouldn't need a Faculty Health Program."

If eyes could cheer, the noise would have been thunderous. In a few moments, several faculty members followed the Chair's example with pet peeves of their own: "Why do seminars have to be scheduled at noon? I can never come at noon." "When will there be a gym for us? They promise it and always cancel it." "Who put exercise machines in the hall? There is no place to change sweaty clothes after using them—so that's a waste of money." Voices rang with the timbre of frustration at a level that indicated that behind these small sources of irritation stood the larger ones: constantly competing demands, time pressure, reduced funding, broken health care system, neglected family, neglected self.

This episode is a pointed enactment of the paradox for programming in faculty health. On the one hand, the steady erosion of health by the conditions of work stays ahead of our efforts to help faculty—not only because those conditions are bigger than we are, but also because those conditions are not addressed by our programming. Nor could they be. Programs for faculty health do not have the power or mandate to bring about the systemic changes that would truly enhance health. So, in the main, our programs occupy the position represented by the book from some years back: *What to Do Until the Messiah Comes* [1].

On the other hand, there is a stubborn fact of life for each of us. As put to us by my strong friend and scholar Walter Wink, we are not responsible for what is done to us, but we are responsible for what we do with what has been done to us [2]. Programming in faculty health is necessary to support the opportunity of each faculty member to provide care for self in the midst of circumstances that are beyond immediate individual control. And part of that support may indeed be providing band aids. Band aids stop the bleeding. They prevent the spread of toxicity. They allow healing. These effects are not nothing.

Programs in faculty health can do more, though. They can discover much about what aspects of institutional culture creates the most personal strain. Programs gather such data by conducting surveys, focus groups, and interviews. But more significantly, seminars and other offerings of faculty health programs stimulate conversations that pinpoint particularly troubling institutional arrangements. This information can be passed on to institutional leaders in a position to make systemic changes.

We in faculty health also hope that methods of self-care learned at the medical center will carry over into the home environment. Being stressed from work hinders high and gratifying functioning at home, but so too does the stress from home reduce high and gratifying functioning at work. Programs in faculty health can break into that vicious circle.

In the end, then, a faculty health program can work at the individual level and at the same time at the institutional, sending messages up the ladder, teaching and supporting all the way down. How to organize a program depends on the politics, funding, and receptiveness to it. Although these factors will affect what topics and formats can be offered, some key elements to consider include the following.

Assessments

Programs in faculty health often become the source for performing assessments of stress in the faculty. The most frequently used standardized measure is the Maslach Burnout Inventory [3]. A study of internal medicine physicians illustrates its use [4]. Focus groups also gauge stress levels and have the advantage of giving information about specific stressors. In order to preserve confidentiality, these are usually facilitated by an outside consultant. Direct interviews with departmental chairs—obviously not confidential—provide significant data about problems noted by those leaders responsible for managing them. General confidential surveys with questions specifically designed for the setting give yet another layer of information. Gathering information from the significant others of faculty yields a perspective not necessarily more "true" than self-report, but certainly different.

Educational Seminars

Workshops and seminars impart skills, knowledge, and motivation for self-care. These can be held at noon for the entire faculty or at a time chosen by a particular group to suit its schedule. Favorite topics concern balancing work with other aspects of life, signs and help for burnout and depression, and cognitive and behavioral strategies for stress management. In-house specialists as well as experts outside the institution give the classes.

Mindfulness-Based Stress Reduction

A frequently used and researched method for eliciting relaxation and a sense of well-being is Mindfulness-Based Stress Reduction (MBSR) (see [5] for an example of a program in MBSR). Based on the work of Jon Kabat-Zinn, training in the method teaches participants to gain access to a resource present in them, but usually overlooked: a non-judgmental awareness of body, mind, heart, and soul [6]. Practitioners contrast this desired mental state with its opposite: mindlessness. Mindlessness is defined as a loss of awareness that results in forgetfulness, separation from the self, and a sense of living mechanically. The typical curriculum for MBSR is 30 hours of instruction spread over eight weeks, but success has also been reached with shorter courses [7].

Psychological Counseling

Employee Assistance Programs typically offer psychological counseling services for all employees, including faculty. But some institutions offer an additional program solely for faculty (sometimes including their families), providing a limited number of sessions with a mental health professional who is outside the institution. This arrangement aims to assure the faculty that not only the content but also the fact of their visit to the consultant will remain confidential.

Reflective Writing

Following upon the recent expansion of interest in the field of narrative medicine and research into the effects of expressive writing, some programs offer retreats in reflective writing, ongoing groups, and one-time experiences [8, 9]. Writing may be done in a careful and structured manner or may be informal, spontaneous, and unedited. Participants report that they experience healing, meaning, clarification, and emotional integration. A further benefit comes from the close relationship that develops among participants as a result of sharing the aspects of their emotional life that is relevant to their work. Without some such forum that facilitates articulation of emotions, many in academic medicine never find a way of sharing themselves at the level of meaning and remain isolated.

Meditation Groups

Herbert Benson, M.D., began in 1968 what has become a long career at Harvard University in research on the benefits of meditation [10]. He has shown direct effects on heartbeat, respiration, oxygen consumption, and skin resistance. Self-reports of the effects include feeling less stressed, more relaxed, and mentally clear. With no necessity to connect it to any religious or spiritual tradition, meditation as a continuous offering at least weekly has several advantages. It provides faculty with training into the relaxation response, an opportunity for building community, and a method that can be easily incorporated into busy days with good results even when applied for only a few moments.

Renewal Groups

Derived from the "circle of trust" proposed by Parker Palmer to explore how to "join soul and role," small groups meet over time with a facilitator for contemplative practice and engaged reflection [11]. The group may look to poetry, reflective writing, music, key stories from participants, fiction, or any other medium that can create access to the deep parts of the self that became committed to service as a medical researcher or clinician. The John P. McGovern, M.D., Center for Health, Humanities, and the Human Spirit at the University of Texas-Houston Medical School has provided a Faculty Renewal Group for the last three years.

Impaired Physician

Many people think that dealing with the impaired physician is synonymous with programming in faculty health, and think alcoholism and drug abuse are synonymous with the impairment of the impaired physician. (The term impairment refers to a hindrance to performing one's work to professional standards.) Actually, programs for faculty health and well-being do not usually direct the program to address impaired physicians, and impairment has come to encompass not only substance abuse, but also disruptive behavior or so-called distressed physicians. Even if a given program for faculty health does not do this work itself, the director of the program for faculty health will likely receive calls about how to deal with someone who appears to have problems with substance abuse or has a pattern of behaving badly. In addition, the director may serve on the advisory committee that does oversee dealing with such troubled faculty.

Although protocols and programs for addressing those with substance abuse have a long history, the label "distressed physician" is of more recent origin and its use is growing. The label comprises an amorphous category of people who have behaviors that disturb others—not occasionally, but chronically. Examples include contentious, threatening, litigious, or intractable behavior. There are two key markers. One is the deviation of the behavior from what is regarded as acceptable by the peer group, and second, the creation of an atmosphere that interferes with the efficient functioning of the staff and the institution. Distressed physicians generally lack skills of self-observation and, when confronted with the behaviors, feel misunderstood and explain the reactions of others as prompted by envy.

Generally, academic medical centers have protocols in place about how to get a distressed physician to a person with expertise in providing testing and assessment, just as they do for faculty with substance abuse. Treatment for both kinds of impairment is often outsourced and includes educational programs as well as behavioral modification. Peer monitoring and loss of privileges are components of the plan. An example of a protocol for addressing problems with distressed physicians is provided by the Tennessee Medical Foundation [12]. Outline of a three-day treatment schedule is one delivered by the Center for Professional Health at Vanderbilt [13]. An approach for all types of impairment is found at Beth Israel Deaconess Medical Center, which refers faculty to the Physicians Health Service, a non-profit corporation of the Massachusetts Medical Society [14].

Web Site

A web site on behalf of faculty health and wellness can itself be an important element of programming. It conveys information about in-house services and seminars and providers, and it also can extend the range and availability of programs. For example, web sites can have links to private instruments for self-assessment of conditions such as depression, alcoholism, or burnout. If scores signal concern, there are resources listed for help both inside and outside the institution. Web sites also can carry educational materials: how to argue successfully, how to gain more balance between time for yourself and time for others, how to help the family during divorce, how to control weight, and how to grieve. For faculty who never attend seminars for health and wellness on-site because of scheduling difficulties or personal reluctance, services through the web site offer a good alternative.

Incentives

Offering incentives of one kind or another has a long pedigree for motivating the slow to move and is a nice reward for those who take action for other reasons. Incentives may be financial or time-off or exercise items (yoga mat, jump rope). An example of the use of financial incentives for wellness activities is given at the University of Arkansas for Medical Sciences [15].

Community

Building relationships and collegiality is a premier way of supporting resilience and protecting against the ravages of stress. Ongoing groups such as a weekly meditation group or monthly renewal group described above offer opportunity for connection in addition to their stated purpose. Serving food after seminars to give people a reason to stay in place and fall into conversation about the day's topic is a more informal way. An innovative method is to organize interest groups—ballroom dancing, photography, reading, and cycling.

Physical Health

Wellness programs may offer Health Risk Assessment—either online using self-report data or on-site with checks on blood pressure, body composition, blood analysis, bone density, and other screens. Classes on nutrition and exercise extend the emphasis. The institution may accomplish extensive programming by contracting with an outside company. An example is the program at the University of Minnesota [16].

Decompression Groups

Clinicians rarely take the opportunity to reflect on, articulate, and integrate the emotions evoked by the death of a patient, by difficult decisions, by strained discussions with families. Researchers as well move from one pressured phase to the next with no formal acknowledgement of the emotions involved in having an experiment go badly, putting all other life aside to complete a grant proposal, having a grant proposal rejected, or having funding cut in the middle of a cycle. Especially because such events and their attached emotions happen routinely, they are not treated as noteworthy. Yet this very chronicity challenges personal systems of renewal and resilience. A meeting whose stated purpose is to decompress from such difficult times puts relief into the schedule. Even a group that meets only once a month can help such matters find their way into informal conversation outside the group.

Ad Hoc

As a program for faculty health and well-being becomes known within the institution, department chairs and individual faculty begin to request specific services for an individual, a team, a department, or the institution. The program itself may seek out the possibility of such specially designed services by interviewing chairs or faculty groups to ascertain their needs and interests. If resources allow for vigorous response to these requests, the offerings of the program and the needs of the faculty become more and more aligned.

Every institution that sponsors a program for faculty health and well-being will toss up many unique elements of programming for us all to consider adding to our own repertoire. But beyond the value of our offerings, there are two further thoughts that frame our programming. The first has to do with the outlook of our customers. Indeed, it has to do with whether they are customers, that is, whether they are in the market for the message we are trying to sell to them.

In my own institution, I have been interviewing the chairs of our 50 departments. Some are far advanced in the realm of self-care and care for their faculty. One keeps a blood pressure cuff on his desk and monitors himself through the day, especially during times when he feels himself to be stressed. As any initiate of stress management knows, learning to notice one's own level of stress is itself an intervention into the spiral upward and takes both commitment and rehearsal. When his blood pressure measures above his target, he sits with eyes closed, does measured, focused breathing for several minutes, gives himself a calming talk, rechecks his pressure to find it normal, and then returns to his work. "If I can't take vacation, at least I can deep breathe," he said. Two or three times a week, he swims—at midnight at the gym on his street. "At that time of night, I own the pool." Then, as the perfect motto for programs in faculty health at our institution, he stated, "If you want to kill cancer, you have to take care of yourself so you live longer and work more." To support the morale and well-being of his faculty, he gives brunches on Saturday mornings for fellowship followed by a quick review of the 18 or so recently published scholarly papers most relevant to his lab. He meets with new junior faculty monthly for career development and to tell them his philosophy of self-care. With senior faculty, he rotates monthly lunches with no agenda. In essence, he runs a faculty health program and welcomes my additions to it.

There is the other end of the spectrum. One chair said, "I have absolutely no issues regarding stress, but you can meet with my faculty." They, of course, turned out not to have any issues, either. Another chair said—and this despite the fact that he carries out every evening a discipline of stress relief through martial arts and meditation, "I am very concerned that promoting the Faculty Health Program might lead people into self-indulgence. People can get confused. This is not a vacation place. It's for hard work, and that is stressful. People should understand that." A third sees relaxation as undermining productivity. He ridiculed the notion of having his faculty members learn relaxing breathing, or even energizing breathing, to reset themselves during the day. "I think it's better just to stay with it, keep at the task, and not see the contrast of that with a relaxed state."

Chairs have enormous influence. They can add to a sometimes prevailing culture that regards self-care as a sign of softness, weakness, or lack of ambition, or they can teach those whom they lead the ethic of self-care articulated above: take care of yourself to take care of your mighty work. And, I would add, to cherish your whole life. Our programs in faculty health, with the help of those leaders who see their value, must address belief systems contrary to this ethic of self-care. Otherwise, no matter how inventive our programming and important our message, we remain shut out by too many.

The second key thought framing our programming came clear in a meeting I held with one of our departments. I asked them to tell me about the challenges in their work that they found particularly stressful. In careful and caring detail, they described the work they do with patients to prepare them for returning home in such different circumstances—now without a jaw or an arm or a stomach. Facing along with the patient the enormity of the change with no place to hide from the difficulties of living in this new way, no place to hide from the patient's downward trajectory—this challenge is the most intense, though there were others.

When I asked what might support them or restore them, I expected them to request something dramatic, something the size of what they had told me. I was ready to advocate for them to get something big. But they wanted a massage paid for by the department, even though the cost is minor, a dollar a minute. "That way, we will feel appreciated." They wanted a faculty retreat like the one a few years back where departmental work was done in a nice setting with a golf course. They wanted a meditation group, the long-promised all-faculty gym, and their elliptical machine put in a more convenient place. I sent them the promise of a high official that the gym is happening. Our Faculty Health Program started a meditation group. And I met with the Department Chair to tell him the other requests, all of which he promptly agreed to.

The point of this story here is not that it may take very little to make a big difference, although that is a point. Rather, the point comes from the prevailing cultural norms that stand behind the response I received one time when I told that story: "Since the Chair was obviously amenable, why didn't the faculty just go to him themselves? Why did the Faculty Health Program need to be involved?"

Because what the faculty needs in order to feel appreciated and renewed is rarely a part of the template for conversation between a Chair and the faculty.

Because what a faculty member needs in order to support his or her own well-being is rarely a part of even the intrapsychic conversation.

Because the topic will not be broached without a forum where it is made legitimate and marked with the institution's imprimatur.

This is the significant mission of programs in faculty health and well-being: to present faculty with a time and space to reflect on their work, its meaning, its importance, and its toll, so that they are moved to value their physical, mental, and spiritual health, and to put forth for themselves what they need in order to live accordingly.

References

1. Fusco, P. (1971) *What to Do Until the Messiah Comes* New York: Collier.
2. Wink, W. (1993) *Engaging the Powers: Discernment and Resistance in a World of Domination* Philadelphia, PA: Fortress Press.
3. Maslach, C., and Jackson, S.E. (1981) The measurement of experienced burnout *Journal of Occupational Behavior* **2(2)**, 99–113.
4. Panagopoulou, E., Montgomery, A., and Benos, A. (2006) Burnout in internal medicine physicians: differences between residents and specialists *European Journal of Internal Medicine* **17(3)**, 195–200.
5. www.umassmed.edu/cfm
6. Kabat-Zinn, J. (1994) *Wherever You Go, There You Are: Mindfulness in Everyday Life* New York: Hyperion.
7. Mackenzie, C.S., Poulin, P.A., and Seidman-Carlson, R. (2006) A brief mindfulness-based stress reduction intervention for nurses and nurse aides *Applied Nursing Research* **19(2)**, 105–109.
8. Charon, R. (2006) *Narrative Medicine: Honoring the Stories of Illness* New York: Oxford University Press.
9. Lepore, S.L., and Smyth, J.M. (2002) *The Writing Cure: How Expressive Writing Promotes Health and Emotional Well-Being* Washington, DC: American Psychological Association Books.
10. Benson, H., and Proctor, W. (2003) *The Breakout Principle—How to Activate the Natural Trigger that Maximizes Creativity, Athletic Performance, Productivity, and Personal Well-Being* New York: Scribner.
11. Palmer, P. (2004) *A Hidden Wholeness: The Journey Toward an Undivided Life* San Francisco, CA: Jossey-Bass.
12. www.e-tmf.org
13. www.mc.vanderbilt.edu/root/vumc.php?site = cph&doc = 4253
14. www.bidmc.harvard.edu/display.asp?node_id = 5683
15. www.ualr.edu/wellness/
16. www1.umn.edu/ohr/wellness

Part VI
Conclusion

Chapter 17
Faculty Health: A New Field of Inquiry and Programming

Thelma Jean Goodrich, Thomas R. Cole, and Ellen R. Gritz

As of this writing, the subject "Faculty Health" does not even appear as a topic in Medline. The components are there (burnout, depression, impairment, etc.), but they have not been brought together under one rubric. Hence, this book is an initial step toward defining Faculty Health as a field of inquiry, programming, and practice. Here, the boundaries of the field have been drawn around the following areas: epidemiology; impairment; psychological issues; gender and generation; racial/ethnic diversity; organizational culture; the faculty life cycle; methods of research and measurement; the ethics of self-care; the humanities as ways of knowing and enhancing; medicine as a calling; and program development.

Theories of and methods for measurement have not been fully developed, but some of the needed measurements concern all of the areas listed above. In addition, we need to assess:

- The institutions' support for faculty health programs
- The effectiveness of programs
- The institutions' support for diverse populations
- Stressors particular to various racial/ethnic, gender, and generational groups
- The distinctive culture of each medical center

As for programming, prevention needs more prominence in the general discussion. Until recently, the programming that existed relevant to faculty health had its strongest focus on intervening for clinicians in situations of impairment. Several shortcomings are worth noting. First, the AMA definition of impairment cited earlier

T.J. Goodrich
Director of Faculty Health, Department of Behavioral Science, The University of Texas M. D. Anderson Cancer Center, Houston, Texas, USA
e-mail: tjgoodrich@mdanderson.org

T.R. Cole
McGovern Center for Health, Humanities, and the Human Spirit, University of Texas-Houston Medical School, Houston, Texas, USA

E.R. Gritz
Department of Behavioral Science, University of Texas M. D. Anderson Cancer Center, Houston, Texas, USA

T.R. Cole et al. (eds.) *Faculty Health in Academic Medicine*,
© Humana Press, a part of Springer Science + Business Media, LLC 2009

in Chapter 3 shows a split between the approaches to mental and physical health. Integrating these services under Faculty Health would model what the health care system should be in American society as a whole. Second, impairment is the end-point on a continuum of declining efficacy of function and should be addressed before the endpoint is reached. Third, little is available for scientific faculty and yet they are also at risk. Especially so are those researchers who may work in relative isolation without ready support and assessment from their associates, and those who are largely dependent on raising their own salaries and research funds.

In a recent article, Kirch and his colleagues describe a project to reinvent the academic medical center along with principles to guide other leaders who want to help their institutions make a fundamental shift [1]. The new design targets organizational structures and processes and at the same time counts on leadership to name, support, and perhaps require self-care for faculty. As our authors have emphasized, self-care as well as institutional care turn largely on recalling the meaning and meaningfulness of the work of the learned professions and aligning the goals and values of individuals, departments, and the institution at the highest levels so that faculty can perform that meaningful work.

It is key that both dimensions occur at once—the recalling to mission and the call to care for the faculty's well-being. If mission alone is emphasized, we will continue to see faculty working too hard to accomplish it. Relational growth and repair need to be balanced with the internal and external push for productivity. In citing economics as the justification for the strong push for productivity, leaders thereby reify "economics." In reality, all decision-making reflects a hierarchy of values. Rather than obscuring their particular priorities, leaders need to acknowledge and question them, give their faculty voice, and make transparency a characteristic of decision-making.

To reiterate, the new field of faculty health is a response to the alarming levels of burnout and demoralization in academic medicine, which cause separation from meaningful work and life. The goal of this emerging field is to study the pressures, measure the consequences, respond with programs, and provide guidelines. At the last session of our conference, participants laid out principles which respond to the need for healthy faculty and healthy institutions:

Table 17.1 Foundations of Faculty Health

1. Academic medical centers create an organizational culture which:
    • Supports individual development, vitality, and resiliency
    • Attends to the challenges and rewards of each stage of the faculty life cycle
    • Respects diversity of gender, race, age, ethnicity, and sexual orientation.
2. Academic medical centers affirm the professional ideals of science and medicine and support faculty in achieving these ideals.
3. Academic medical centers attend to the mutual adjustment between individual goals and institutional goals.
4. Institutional leaders clarify and question the assumptions and priorities behind what are regarded as financial necessities and engage faculty in open discussions.

(continued)

Table 17.1 (continued)

5. Academic medical centers develop a Faculty Health Program as one component of support for faculty vitality and productivity.
6. Academic medical centers emphasize that care for others and care for the self are inextricably linked and that success does not require sacrifice of human wholeness. Leadership publicizes the importance of self-care and provides resources that promote it.
7. Academic medical centers promote collaboration, mentorship, transparency, and mindfulness in professional relationships and enforce these practices as requirements for membership in the faculty.
8. Academic medical centers require a high level of skill in interpersonal relationships as an essential criterion for people in positions of leadership.

We offer these guidelines and the foregoing ideas about faculty health for consideration by all concerned. We look forward to the development of this new field which is vital to the future of academic medicine.

Reference

1. Kirch, D.G., Grigsby, K., Zolko, W.W., et al. (2005) Reinventing the academic health center *Academic Medicine* **80**, 980–989.

Afterword

At The University of Texas M. D. Anderson Cancer Center, ten years ago we held a series of focus groups and developed a statement of institutional values: caring, integrity, and discovery. The process brought out many issues, and the most important was agreement that caring was critical in our relations both with our patients and with each other.

So for the first of our three values, the values statement reads as follows:

Caring – by our words and actions, we create a caring environment for everyone.
We are sensitive to the concerns of our patients and our coworkers.
We are respectful and courteous to each other at all times.
We promote and reward teamwork and inclusiveness.

Note that all three of the actions (behaviors) that describe caring describe how we will treat each other.

As this book so clearly points out, the faculty at academic medical centers spend far more time and effort on caring for patients and their families than for each other. For us, the totally unexpected suicide of a revered surgeon was a wake-up call to action. Faculty leaders undertook a mobilization effort, which resulted in a number of initiatives that remain active eight years later:

A Faculty Health Committee charged with making recommendations to improve the well-being of our faculty

Recruitment of a new faculty member with full-time responsibilities as Director of Faculty Health

A three-part faculty health program focusing on prevention, intervention, and response to catastrophic events

Expansion of the Employee Assistance Program in our Department of Employee Health

Focus groups, lectures, seminars, and courses on a wide variety of issues related to quality of life and support of faculty and their families

A two-day workshop on faculty health, which resulted in the publication of this path-breaking book

The challenges that an academic clinician and biomedical researcher (often the same person) face today are well known. The institutional and personal missions, which focus on improving knowledge of biology and disease and delivery of care, must be blended with the need for funding and the needs of the spouse and family. Overarching these pulls and pushes for the faculty member's attention is the overreaching need in every human being's life for *meaning*.

The problem is that meaning is measured not only by the individual, but also by many with whom the individual comes into contact. For academicians in the medical field, this includes (beyond family) department chairs and deans; peers and collaborators in the clinic and in science; editors of journals and planners of national meetings; trainees; and, of course, our patients. The tugs on our allegiances can become overwhelming challenges, and the tools which we use to cope with these challenges are not priority areas of research and intervention.

This book is written by academicians who care deeply about these important issues. The scholarly chapters bring us up to date on what we know, and what we need to learn about through more research; and what we are doing well, and what we need to do better.

I am convinced that the interventions taken at M. D. Anderson Cancer Center, which are described in Chapters 1, 8, 13, 15, and 16 have had a tremendous impact on the well-being of our faculty and, therefore, on the effectiveness of our patient care and research. I agree with the thoughtful recommendations of the Editors. And I thank all of the contributors to this volume for helping me and others to focus on what we can do to improve the quality of life and productivity of faculty, in the highly charged environment in which they work in today's academic medical centers.

John Mendelsohn, M.D.
President
The University of Texas M. D. Anderson Cancer Center (M. D. Anderson)
Houston, Texas

Appendix
Foundations of Faculty Health: A Consensus Statement of Editors and Authors

---

Foundations of Faculty Health

---

1. Academic medical centers create an organizational culture which:
   - Supports individual development, vitality, and resiliency
   - Attends to the challenges and rewards of each stage of the faculty life cycle
   - Respects diversity of gender, race, age, ethnicity, and sexual orientation.
2. Academic medical centers affirm the professional ideals of science and medicine and support faculty in achieving these ideals.
3. Academic medical centers attend to the mutual adjustment between individual goals and institutional goals.
4. Institutional leaders clarify and question the assumptions and priorities behind what are regarded as financial necessities and engage faculty in open discussions.
5. Academic medical centers develop a Faculty Health Program as one component of support for faculty vitality and productivity.
6. Academic medical centers emphasize that care for others and care for the self are inextricably linked and that success does not require sacrifice of human wholeness. Leadership publicizes the importance of self-care and provides resources that promote it.
7. Academic medical centers promote collaboration, mentorship, transparency, and mindfulness in professional relationships and enforce these practices as requirements for membership in the faculty.
8. Academic medical centers require a high level of skill in interpersonal relationships as an essential criterion for people in positions of leadership.

Index

Printed in the United States of America